The Advanced Smart Grid

Edge Power Driving Sustainability

Second Edition

For a listing of recent titles in the
Artech House Power Engineering Library,
turn to the back of this book.

The Advanced Smart Grid

Edge Power Driving Sustainability

Second Edition

Andres Carvallo
John Cooper

ARTECH
HOUSE

BOSTON | LONDON
artechhouse.com

Library of Congress Cataloging-in-Publication Data
A catalog record for this book is available from the U.S. Library of Congress.

British Library Cataloguing in Publication Data
A catalog record for this book is available from the British Library.

ISBN-13: 978-1-60807-963-6

Cover design by John Gomes

To Angela, Alexandra, Andres Josephe, and Austin Theodore.
—AC

To Barbette, Blake, and Wesley.
–JC

Contents

CHAPTER 5

Envisioning and Designing Smart Grid 2.0   105

CHAPTER 6

Today's Smart Grid   129

CHAPTER 7

Foreword by Jon Wellinghoff

In my 40 years in the energy industry, we have seen innovation in the areas of generation technologies, renewable system production, smart grid implementation, storage, efficiency, demand response, and infrastructure security. The last 10 years though, have seen an acceleration, disruption, and adoption pace driven by unprecedented advances in policy, utility restructuring, technological innovation, and more active consumerism. One of the foundational transformations to help us reach the collective goals of less carbon, lower total cost of ownership, more efficiency, more consumer choice, more services, and more quality has been the advancement of smart grid technologies.

The Advanced Smart Grid: Edge Power Driving Sustainability by Andres Carvallo and John Cooper is a seminal work describing in great detail the vision, rationale, and journey of Austin Energy, the eighth largest public power utility in the nation, toward delivering cleaner, affordable, and more reliable power, coupled with superior customer service. This book also recounts the true story of the Pecan Street Project, now Pecan Street Inc., an initiative to build a smart grid laboratory within a living community to accelerate energy technology innovation and customer adoption. It does so while also giving us first the context of how our electric system evolved from the Edison and Tesla days, to the Texas vision of Pecan Street and, as they say, everything is bigger in Texas, which is where we should move forward to in the years to come.

I met Andres Carvallo in the halls of NARUC conferences in 2008, and our conversations on demand response and smart grid since then have always been ones of mutual fascination. Clearly, the work done by him and many others at Austin Energy have contributed in many ways to the shape of the electric industry today. Their early work built a much needed can-do attitude on how utilities must learn to balance keeping the lights on and energy costs low while striving to reduce carbon and become more efficient and sustainable enterprises (e.g., empowering customers, optimizing energy service delivery, and reducing water use).

From my days on the staff of the Nevada Public Utility Commission and later as a commissioner and then chair at the Federal Energy Regulatory Commission (FERC), Austin Energy always stood out to me as one of those early adopters leading national trends in policy and technology innovation. I can honestly say that following the journey of Austin Energy provides one with a stellar example of what it can be when it comes to a cleaner and more efficient power grid that strives to integrate emerging distributed energy resources such as solar, storage, and demand response into the distribution grid.

Between 2003 and 2010, Andres and his colleagues at Austin Energy turned the following definition of smart grid (SG1) into reality: "The smart grid is the integration of an electric grid, a communications network, software, and hardware to monitor, control, and manage the creation, distribution, storage and consumption of energy. The smart grid of the future will be distributed, it will be interactive, it will be self-healing, and it will communicate with every device."

They then progressed to the more ambitious definition of an advanced smart grid (SG2): "An advanced smart grid enables the seamless integration of utility infrastructure, with buildings, homes, electric vehicles, distributed generation, energy storage, and smart devices to increase grid reliability, energy efficiency, renewable energy use, and customer satisfaction, while reducing capital and operating costs."

In this book, one sees those definitions unfold as real live systems in an ever-advancing architecture that is methodically implemented by the Austin team. Andres Carvallo and John Cooper do a masterful job conveying to utility industry practitioners, regulators, and consumers alike a great episode in our country's energy evolution that should not be missed.

Jon Wellinghoff
Former chairman of the
Federal Energy Regulatory Commission
and partner at Stoel Rives

Foreword by Larry Weis

Andres Carvallo and John Cooper have written a thought-provoking and insightful book on smart grid and, in particular, its potential from leveraging the technology-rich world that we live in. In the 1980s when I was in distribution and construction management at a utility in the Seattle area, there were products being marketed with some built-in intelligence (e.g., electronic reclosers for circuits). Similarly, there were plenty of discussions in the industry about using information systems and communications technology along with our power equipment to help us implement "distribution automation" pervasively. So the ideas of automation had existed for a long time, but technology needed to catch up, become easier to use and manage, and become much more affordable. Over the last 10 years, technology solutions for automation have caught up, and *The Advanced Smart Grid: Edge Power Driving Sustainability* shares and embraces the opportunities that exist to use technology to change how we think of and operate the bulk electric systems and, in particular, how we deliver and manage products and services in partnership with our consumers.

When I came to Austin Energy in 2010, I knew that we were on the cutting edge of innovation. However, our delivery network was still behind. As part of our smart grid program, this year we launched, with the help of our partners, a brand new advanced distribution automation system (ADMS). We believe that our ADMS is unparalleled for large utility operations in the United States and that it will leapfrog us over many other utilities for years to come. Furthermore, we are advancing many of the ideas in this book into our thinking, specifically those about how to develop and leverage microgrids and those on how to advance the state of the art in home energy management.

While at a practical level we are still simply seen as providing the product of reliable and affordable energy to our customers, we have a great opportunity to further experiment in our wonderful urban service area with a lot of the ideas discussed in this book. We strive to be cleaner by deploying more renewables, to use more technology to automate, and to integrate electric vehicles and energy storage into the steady state physics of the grid seamlessly and affordably. I know with certainty that many big challenges lie ahead, and that while I do not get as much time to contemplate the future as my team does, Smart Grid evolution and the range of possible applications are intriguing and exciting things to think about.

Thank you Andres for your past work, for the time that we have spent collaborating on our vision of the future, for the fresh thoughts, and for the ideas presented in *The Advanced Smart Grid*.

Larry Weis
General manager and
CEO of Austin Energy,
Austin, Texas

Preface

Five years have gone by fast. Five years ago, we began work on *The Advanced Smart Grid: Edge Power Driving Sustainability* manuscript, fresh off our experience with Austin Energy and the Pecan Street Project. In the ensuing years, we've both enjoyed active engagement with our industry and passive observation, as the twin threads of grid optimization and edge power developed, closely tracking our observations in the first edition.

By early 2014, we noted that while our work had stood the test of time since its publication in 2011, it begged for an update in certain spots. So, we began reviewing our book for the second edition; it was something of a surprise but very gratifying to read our words set down in 2010 and 2011 and see that they remained timely three to four years later. In fact, the reasoning behind the advanced smart grid (Chapter 1), the fundamental components of the advanced smart grid (Chapter 2), smart convergence with other industries (Chapter 3), the case study of Austin Energy (Chapter 4), and the case study of the Pecan Street Project (Chapter 5) are timeless. The changes in the second edition are summarized as follows:

- Chapters 1–5: When we sat down to write, we found that some minor changes to Chapters 1–5 were needed, primarily edits to lend more precision where we saw room for improvement. Also, we moved any language that referred to specific dates or events to Chapter 6 to set up future editions.

- Chapter 6: The second half of Chapter 6, which covered smart grid in 2010, was quite dated, presenting us with a grand challenge; we spent months in consultation, struggling with how to cover all that had occurred from 2011 when the first edition was published up to the present. In the end, we decided that while grid optimization had progressed steadily, and would continue to do so, the compelling story was with our subtitle, Edge Power Driving Sustainability. It would be far more noteworthy to focus our attention in the second edition on distributed energy resources (DERs),so we devoted our efforts to a completely new second half focused on the disruptive impact of DER.

- Chapter 7: Our description of Smart Grid 3.0 (SG3) remained quite relevant, less focused on a snapshot of time like Chapter 6. We identified some gaps based on our maturing vision of the future, in particular updating the consumer maturity model; adding sections on electric vehicles (EVs) and energy storage (ES), positive energy buildings (PEBs), smart building/smart grid convergence and on the Internet of things (IoT); as well as outlining likely industry structure for the SG3 era.

- Glossary: Finally, we added a glossary with well over 100 acronyms and terms, acknowledging the tremendous challenge for a student of smart energy and smart grid to grasp this complex industry.

We anticipate publishing future editions as we track the progress of advanced smart grid optimization and the rising edge power economy, including the high drama of DER disruption and integration. With each edition, we'll improve our work where we can. In fact, it will likely be necessary to publish more frequent updates in this highly dynamic environment. It is a privilege and honor to find ourselves in the fortunate position to chart the progress of change in this vital industry, and we look forward to the fascinating journey of co-creation, commentary, and analysis.

Opportunity Meets Planning

The illustrious inventor Thomas Edison once said, "Good fortune is what happens when opportunity meets with planning." Throughout our careers as corporate athletes and entrepreneurs, we've examined complexity and explored ways to simplify it, ranging from addressing complex processes and concepts in technology and engineering, to finding the kernels of truth in government research and executive briefings in the legislative and regulatory arena. Lacking precedent, we worked through the complexities of utility processes, applications and systems, operational technology, information technology (IT), telecommunications, and power engineering at Austin Energy and identified a path to simplify and innovate to build the very first smart grid in the United States and then refined our vision with our work in the Pecan Street Project.

This book describes in detail our experience in designing and building the very first smart grid in the United States at Austin Energy—what we now call a first-generation smart grid, or Smart Grid 1.0 (SG1)—and in helping to design an energy internet at the Pecan Street Project (now called Pecan Street, Inc.) to evolve Austin Energy into a Smart Grid 2.0 (SG2) utility, which we've elaborated on to create our advanced smart grid vision. In these pages, we start with the vision that sprang from those unique experiences from 2003 to 2010; then we go back to share our local perspective in Austin (it is the "city of ideas," after all); finally, we share our observations of current events and our vision for the future, explaining what lies ahead for our industry and society.

Necessarily, we focus our story on the new power engineering concepts needed to drive this transition to a more rational approach to designing and operating an advanced smart grid—look for the "Power Engineering Concept Briefs" throughout. We also include use cases where it makes sense to communicate concepts more clearly. This is a highly complex industry on a good day, and as we set off on this fundamental industry and business transformation, it will only get more complicated. It pays to roll up your sleeves and get down in the weeds, as they say.

The remainder of this preface provides an overview of our approach, followed by a public acknowledgment of the many people who have helped us to get to where we are. We hope you enjoy this book and let us know what you think. We live in an interactive world now, and this will certainly be an iterative process. Together, we'll get this transformation right. We have to.

Chapter 1: The Inevitable Emergence of the Smart Grid

In Chapter 1, we draw a distinction between smart grids as they are described, designed, and built—what we term first-generation smart grids (SG1)—and second-generation smart grids, or advanced smart grids (SG2), which begin to emerge as a new understanding takes hold in this industry in line with the vision we have described in this book. First-generation smart grids start with an application, such as advanced metering infrastructure (AMI), and build a smart grid incrementally by adding more applications over time. In contrast, advanced smart grids start with smart grid architecture as part of a deliberate design that includes integrated Internet protocol (IP) network design, thereby positioning the smart grid to support any variety of applications as they become necessary or available.

We describe the electric grid as the most important of all the infrastructures we depend upon in our modern economy and society, going so far as to insert electricity at the base of Maslow's hierarchy of needs. We assert that it is inevitable that the grid will be upgraded to become an advanced smart grid because it is the quintessential infrastructure, but also because technology evolves to empower individuals at the edge over time; the electricity industry will follow similar trend lines described by evolution in the IT and telecommunications industries.

Today, the grid is brittle and challenged—in need of a new architecture. The way forward will be through a new design and overhaul to make it more resilient and even more robust. As the number of connected devices increases dramatically, the level of complexity in the grid will rise to the point where automated protocols are needed to maintain stability—and an Internet design will be required to enable the transfer of very large amounts of data and to ensure that the grid remains functional and continues to supply us with the power on which we are so dependent every minute of every day.

Chapter 2: The Rationale for an Advanced Smart Grid

In Chapter 2, we drill down to explore the impact that extending intelligence to the edge will have on utility network architecture, business processes, and organizational structure. The distributed control system (DCS) has traditionally been generation-oriented, in so much as the management system was comprised of a software program running on a dedicated computer providing directions to automatic generation controllers at one or more central generation units (i.e., power plants) to manage all the switches, boilers, and other devices through control systems, throttling the turbines up and down to maintain grid voltage levels within a specified tight band (60 Hz in the United States).

The rationale for an advanced smart grid is not hard to understand. In a sense, progress in grid management has been about gaining greater efficiencies through better control and better information. Pushing intelligence out into the grid, traditionally accomplished through independent appliances, applications, and networks, will become the purview of an integrated advanced smart grid. For that to happen, however, utility business processes such as annual departmental budget building must also be addressed. The shift from an industrial approach of long product life cycles, to a more IT-oriented environment implies not only managing a more dynamic product market, but also integrating with an IT department and considering the impacts on the greater ecosystem of interconnected devices and data.

Power Engineering Concept Briefs

The advanced smart grid will require departmental managers to coordinate like never before with IT staff on security, network strategy, interconnectivity, and network integration. Common databases will drive applications, minimizing the need for complex integration projects. The benefits and implications of a system rationale will be fully explored in Chapter 2, as planning for a new era of robust digital networks gets under way, where everything connected to the smart grid has become smart in its own way. The advance of technology will inevitably encroach upon traditional utility domains, bringing changes to the traditional utility business model and to the way a power engineer approaches grid management.

Chapter 3: Smart Convergence

Chapter 3 describes the ongoing opportunity for change that two megatrends present to every infrastructure that supports our modern ecosystem, from electricity, to telephony, to the Internet, to water, gas, and transportation. What are the two megatrends? First, ongoing analog-to-digital transitions are based on advances in digital technology driven by Moore's law. Digitization brings faster, cheaper, more powerful computing capabilities to edge devices that transform the potential of infrastructure design and operations. Complementing that trend is the second megatrend, advances in telecommunications and networking, driven by Metcalfe's law. As more wired and wireless technologies become available, infrastructures gain a tool to add digital devices and modernize their infrastructure operations. Furthermore, as all these infrastructures begin to transform themselves, they converge on each other, offering still more synergy.

Complementing this convergence of infrastructures is a convergence of business practices from other industries onto the electric utility infrastructure, perhaps best exemplified by the addition of warehouses and greater edge controls to the electric supply chain. Lacking effective storage or demand-side management options, the electric utility supply chain developed historically with an overwhelming reliance on supply-side solutions to keep voltage and VAR levels in harmony. Keeping the grid in balance is the overriding goal of the utility controller. However, with the addition of energy storage and maturing of demand response (DR), electric utility operators will see storage and demand-side management as valid alternatives to generation control—valuable new tools to keep the grid in balance.

Power Engineering Concept Briefs

This chapter has perhaps the most comprehensive set of concept briefs of all the chapters, describing in detail how the convergence between infrastructures and the convergence of new concepts will work at the engineering level to bring changes to the grid. From a description of how to build a "thin" smart grid wireless IP network, to the detailed discussion of incorporating new digital tools into grid management solutions, these sections provide an engineering drilldown for the technically minded.

Chapter 4: SG1 Emerges

Chapter 4 tells the story of how the very first comprehensive, utility-wide first-generation smart grid came to be built in Austin, Texas, at Austin Energy, the city-owned electric utility that now serves over 420,000 residential, commercial, and industrial customers. The lessons learned in the smart grid journey described in this case study are fundamental to understanding the concepts in this book, which derive from the successes and lessons learned in Austin, in Texas, and in the United States from 2003 to the present.

The chapter describes the initial assessment, efforts to realign IT processes and gain organizational buy-in, and communication of a new, more comprehensive vision. Key to the transition were institutional tools such as a technology governance plan, a technology leadership team, customer steering committees, a project management office (PMO), an enterprise architecture council, a technology security council, a disaster recovery Council, and an enterprise data council, which together took the utility on an evolutionary journey from technical anarchy to standardization, increasing productivity and instituting proactive control.

Power Engineering Concept Briefs

Any advanced smart grid project needs two things above all else. First, it needs project funding, which can be found in part from system rationalization that eradicates wasteful IT spending and frees up cash to fund strategic initiatives like an advanced smart grid project. Second, organizational buy-in is needed from other departments, achieved by improved communication, better service, and cross-functional experiences.

Chapter 5: Envisioning and Designing SG2

In Chapter 5, we tell the story of the Pecan Street Project, a unique community project in Austin, Texas, whose goal was to envision and design an energy internet, essentially an advanced smart grid, launched in late 2008 and continuing on to 2013 as an ARRA-funded U.S. Department of Energy (DOE) Smart Grid Demonstration Project, then as an independent research company. The project was informed by the experiences at Austin Energy from 2003 to 2009 in building a pioneer first-generation smart grid. Weekly brainstorming over 25 weeks allowed the diverse teams to evaluate and imagine the next generation of the smart grid, which proved immeasurably valuable to the authors, offering keen insights to help define the essentials of the advanced smart grid vision outlined in this book.

The Pecan Street Project, named for a main cross street in downtown Austin, began in 2008 with an idea for clean energy–led economic development. By harnessing community input, the project founders hoped to capture the imagination of the nation and steer clean energy companies to locate and grow in Austin and central Texas, creating a new focal point for economic development to complement the semiconductor and Internet business foundations of Austin's new age economy. While the jury is still out on the success of that primary goal, the 200-person team did succeed in laying the groundwork plans for a new clean energy ecosystem, whose central tenet is the importance of integrating the water and transportation infrastructure with the power infrastructure. Beyond that level of integration, the project also emphasized the importance of integrating the community into the decision and planning process, since so much of the DER equation depends on an informed and motivated citizenry to move beyond niche applications into mainstream adoption. Throughout this book, DER is a term that includes DR, distributed generation (DG), electric vehicles (EV), energy storage (ES), and appliances and systems with built-in energy-efficiency (EE), collectively the new energy technologies that comprise both the supply and demand sides of "edge power"; the Pecan Street Project discussion expanded the term beyond supply-side technologies to include DR and EE.

Four key elements emerged in the storyline of Phase One of the Pecan Street Project: first, the need to integrate technology, specifically the emerging DER technologies, but also water technologies; second, the emerging need for a new business model for utilities to replace the 100-year-old model of distributing commodity kilowatt-hours; third, the need to integrate and motivate the energy consumer into the energy ecosystem; and fourth, the need to channel regulatory and legislative policy to accommodate all the necessary changes (all these have become key trend lines in the 2014 energy debate documented in Chapter 6).

In what we then called Phase Two, the nonprofit Pecan Street Project organization used DOE grant money and local funding in a three-year study of an energy Internet neighborhood at a new urban style neighborhood located on the site of the old Mueller Airport in central Austin. Pecan Street has since rebranded itself as a research network linking a neighborhood with intense DER penetration with UT's mainframe computers.

Power Engineering Concept Briefs

The Pecan Street Architectural Framework, and all the other elements of the Pecan Street Project process offer tremendous lesson plans from the power engineering perspective, discussed in detail in this section.

Chapter 6: The National Perspective on Smart Grid

In Chapter 6, we pause from describing our local journey on the path to a smart grid in Austin at Austin Energy and the Pecan Street Project to set the context at the national level. While it remains critical to understand the changes that must take place inside a utility and within a community, it's also important to understand the context at the regional and national level. We look at the origins of the smart grid movement in the United States and globally, and then focus on three principal stakeholder groups in Smart Grid: national and state-level governmental activity, industry standards, and industry stakeholder groups. The second half of Chapter 6 is comprised of a review of the disruptive impacts of DER on grid modernization—a critical factor in the advance of SG2—in three sections: Technology Innovation, Edge Power Policy Trends, and Business Model Trends.

Chapter 7: Fast-Forward to SG3

In Chapter 7, we take time to contemplate the future, in a world where the advanced smart grid has become the accepted norm, and we begin to realize the system benefits of a rationalized regional smart grid. We discuss the standards and templates that are only now being contemplated but that when implemented will guide the deployment of advanced smart grids. We explore the different strategies and capabilities that will be needed to overcome challenges and obstacles of growing complexity that will appear along the road to an advanced smart grid.

We also examine anew the planning and design for advanced smart grids. We focus on the role of a smart grid architecture framework (SGAF), which provides a cookbook of how-to recipes for grid designers. We discuss the role of a smart grid optimization engine (SGOE), which provides a mechanism for planning and operation of an advanced smart grid using real-time, event-driven decisions in a data-rich environment. And we explore the integration of distributed elements into the SGAF and SGOE, where future planners and operators will use new technologies like DR management system (DRMS), DER management system (DERMS), and intelligent load management (ILM) software to plan and operate an advanced smart grid with millions of devices, far beyond what they must cope with today.

To conclude our journey, we look briefly at SG3, where a vision of a clean, linked future involves such things as peer-to-peer energy trading (P2PET), now referred to as transactive energy; energy roaming and EVs; ES; virtual power plants (VPPs); and microgrids. When pervasive IP networking and computing and grid power become commingled with abundant information and edge-based DER, new forms of energy trading and consumption will become possible, similar to the flow of information in the maturing Internet economy. We have termed this ambitious future SG3.

We also include use cases in Chapter 7 that describe three real-life scenarios and the different features of the advanced smart grid that will ultimately enable the SG3 vision.

Acknowledgments

We start our acknowldgements with our families, whom we love deeply and who have been denied our company over the innumerable weekends and long evenings while we've been busy writing the first and second editions of this book. We'd like to acknowledge as well a few of the many friends, mentors, bosses, colleagues, and associates who taught us, connected the dots, were there to brainstorm with us, and, finally, inspired us with their insights over the years (and our apologies to those left off this list; there were so many more, too many to acknowledge individually):

At AEP: Mike Thomason, Mike Babin, Charles Patton, and Stuart Solomon; at Austin Energy: Larry Weis, Juan Garza, Roger Duncan, Elaine Hart, Bob Kahn, Michael McCluskey, Cheryl Mele, Chris Kirksey, Kerry Overton, Ed Clark, John Baker, David Wood, Mark, Dreyfus, Brian Davison, Andy Perny, John Tempesta, Al Sarria, Keith Rabun, Debbie Starr, Jerry Hernandez, J. J. Guitierrez, Mike La-Marre, and Karl Popham; at Borland: Philippe Kahn, Rob Dickerson, Doug Antone, and Steve Schiro; at Cisco: Laura Ipsen, Paul DeMartini, and J. D. Stanley; at City of Austin: Will Wynn and Rudy Garza; at DEC: Enrico Pesatori, Harry Copperman, Ada Holian, Rafael Pineiro, and Nicky Lecaroz; at Direct Energy: David Dollihite and Kelly Hayes; at Gartner: Zarko Sumic; at Dell: Dwight Moore; at GE: Bob Gilligan, Mark Hurra, Steve Richards, Kerry Evans, and Mike Carlson; at GridBOT: Richard Donnelly; at GridWise Alliance: Katherine Hamilton; at IBM: Steve Mills, John Soyring, Guido Bartels, Brad Gammons, Allan Schurr, Scott Winters, Paul Williamson, Mike Francese, and Amy Thomas-Gerling; at IDC Energy Insights: Rick Nicholson; at iMark.com: Brian Magierski; at KU: Stuart Bell, Perry Alexander, Gary Minden, and Larry Weatherley; at Lighthouse Solar: Stan Pipkin; at Lotus: Jim Manzi and James Henry; at Microsoft: Bill Gates, Steve Ballmer, Scott Oki, Adrian King, Greg Diaz, Gabe Newell, Cameron Myhrvold, Rob Horwitz, and Joseph Mouhanna; at New Brunswick Power: Brad Wasson, Michel Losier and the RASD team; at NIST: George Arnold; at Northeast Energy Partners: Erich Landis; at OGE: Chris Greenwell; at Pecan Street Project: Brewster McCracken, Colin Rowan, and Jim Marston; at Phillips Electronics: Mike McTighe, Osmo Hautenan, Terry Vega, Delia Schneider, and Paul Murdock; at SCO: Larry Michels, Doug Michels, Sam Spadafora, Xavier Montserrat, Don Morrison, Mike Foster, and David Bernstein; at Sharp Labs: Carl Mansfield; at Siemens: Pierre Mullin, Sonya Hull, Tugcan Sahin, Daniel Shereck, Tim Gibson, Jon Horsman and Jack Joyce; and at UTC: Bill Moroney; at VENewNet: Tom Dickey; at UT: Ed Anderson, Mack Grady, and Ross Baldick; and at Xtreme Power: Michael Breen.

Introduction by Thomas M. Siebel

In 1976, Daniel Bell, a leading sociologist from Harvard, published a book entitled *The Coming of Post-Industrial Society*, in which he predicted the advent of the Information Age. Years before the invention of the Internet as we know it today, the minicomputer, the personal computer, and the cell phone, Bell predicted that information and communications technology would affect a fundamental change in the structure of the global economy; a change on the order of magnitude of the Industrial Revolution [1].

The information age would portend the preeminence of the "knowledge worker" and result in the emergence and growth of the information technology industry driving fundamental and ubiquitous changes in the ways we work, communicate, and recreate.

With the genesis of the smart grid, Bell's predictions meet the business value chain of power generation, transmission, distribution, and consumption.

As much as $2 trillion is being invested this decade to upgrade the power infrastructure globally including adding sensors to make the devices in the power grid remotely machine addressable. These newly sensored devices include smart meters, thermostats, home appliances and HVAC equipment, factory equipment and machinery, transformers, substations, distribution feeders, and power generation and control componentry. This is the physical infrastructure of the smart grid [2].

Nascent at the beginning of this century, the smart grid is advancing apace. As of 2014, nearly 400 million smart meters have been installed globally. That number will more than double in the coming decade [3]. Representing a fraction of the sensors on the grid infrastructure, the smart meter installations serve as a proxy for the penetration and growth rate of the advanced smart grid. The National Academy of Engineers identified the electric grid as the most significant scientific achievement of the twentieth century [4]. The advanced smart grid will be the largest and most complex machine ever conceived and will likely prove one of the most significant scientific achievements of the twenty-first century.

Andres Carvallo and John Cooper's newest release of *The Advanced Smart Grid: Edge Power Driving Sustainability* manifests the authors' deep industry acumen and unique ability to identify and frame the underlying trends, transformative opportunities, and technological advances for energy systems.

The authors' prescient vision, deep understanding, and clear articulation of the power of the advanced smart grid is compelling and convincing. Moore's law meets

Metcalf's law—their view of the smart grid as a fully connected, sensored internetwork is an epiphany. Their grasp of the history and developments in information and communication technology portrays the advanced smart grid as an imminent certainty, not a future possibility. The authors masterfully describe how consumer expectations, foundational technological trends, and innovations in technology interact to enable the development of the communications network, software, and hardware of the advanced smart grid.

The truth is that the "smart" sensored devices in themselves provide little utility. They simply provide the capability to remotely sense and/or change a device's state. For example, is the device operative or inoperative? If operative, at what temperature, voltage, or amperage? It might allow us to know the amount of energy that the device has consumed or recorded over some period of time or is consuming in real time.

Collectively, these devices generate massive amounts of information—an increase of six orders of magnitude. As utilities adapt from managing a relatively small number of noncommunicating devices to connecting hundreds of millions of diverse sensored devices, data volumes will expand exponentially. To effectively aggregate and manage this influx of data, utilities require new technologies to integrate, process, apply analytics and machine learning, and intuitively visualize the data and analytic results in a way that drives business outcomes through a common data- and intelligence-driven smart grid operating system.

The authors describe this required technology as a smart grid optimization engine (SGOE). With an SGOE, utilities can control the multitude of devices that constitute the smart grid. By leveraging recent advancements in information technology, including elastic cloud computing and the sciences of big data, machine learning, and emerging social human-computer interaction models, the sum of *all* information generated by smart grid infrastructure is collected, aggregated, correlated, and scientifically analyzed in real time in a single, federated platform. The SGOE allows utilities to operate in a complex, dynamic environment to enable real-time and predictive modeling to realize self-healing, automated, and efficient operations.

Analytics enabled by the advanced smart grid provide the capability for efficiencies across the energy value chain, including real-time pricing signals to energy consumers, management of sophisticated energy efficiency and demand response programs, conservation of energy use, reduction of fuel necessary to power the grid, real-time reconfiguration of the power network around points of failure, instantaneous recovery from power interruptions, accurate prediction of load, efficient management of distributed generation capacity, rapid recovery from damage inflicted by weather events and system failures, significant conservation of energy, reduction of fuel needed to power the grid, and substantial reduction in adverse environmental impact.

SGOE solutions, as described by the authors, integrate massive amounts of disparate data, apply sophisticated multilayered analytics, and provide highly usable portals that generate actionable real-time insights. They have the power to integrate billions of rows of data from scores of source systems and exceed processing speeds of 6.5 billion transactions an hour. These solutions are being proven at scale today addressing more than 52 million smart meters globally [5]. The SGOE provides utilities with end-to-end system visibility across supply-side and demand-side smart grid operations.

SGOE solutions enable grid operators to realize dramatic advances in safety, reliability, cost efficiency, and environmental benefit by correlating and analyzing all of the dynamics and interactions associated with the end-to-end power infrastructure as a fully interconnected sensored network including current and predicted demand, consumption, EV load, distributed generation capacity, technical and non-technical losses, weather reports and forecasts, and generation capacity holistically across the entire value chain. Leading utilities including Enel [6] and Exelon [7] are driving innovation by applying SGOE solutions that combine the sciences of cloud-scale computing, advanced smart grid analytics, and machine learning to the benefit of their communities, consumers, stakeholders, and the environment. These utilities are at the vanguard; others will follow.

Every decision maker who manages, engineers, and transacts with, plans, or regulates our energy systems should read this thoughtful book. The authors' vision is refreshingly clear, identifying the key fundamentals that underlie the tremendous challenges and opportunities of the advanced smart grid. This book incisively captures the inevitability of the advanced smart grid and the many exciting opportunities it affords.

Thomas M. Siebel
Chairman and Chief Executive Officer of C3 Energy
Former Chairman and Chief Executive Officer of Siebel Systems
Fellow, American Academy of Arts and Sciences
Board Member for the Stanford University College of Engineering,
the University of Illinois College of Engineering, and
the University of California
at Berkeley College of Engineering.

Endnotes

[1]   Bell, Daniel. *The Coming of Post-Industrial Society: A Venture in Social Forecasting.* New York: Basic Books, 1976.

[2]   Electric Power Research Institute. *Estimating the Costs and Benefits of the Smart Grid: A Preliminary Estimate of the Investment Requirements and the Resultant Benefits of a Fully Functioning Smart Grid.* EPRI, Palo Alto, CA: 2011. 1022519

[3]   Navigant Research. *Smart Electric Meters, Advanced Metering Infrastructure, and Meter Communications: Global Market Analysis and Forecasts.* Chicago: 3Q 2014.

[4]   National Academy of Engineering. "National Academy of Engineering Reveals Top Engineering Impacts of the 20th Century: Electrification Cited as Most Important." Web. February 22, 2000. http://www.nationalacademies.org/greatachievements/Feb22Release.PDF.

[5]   C3 Energy. "C3 Energy Reports Record Midyear Performance." Web. November 11, 2014. http://c3energy.com/community-newsroom-readmore?nid=230

[6]   Engerati. "Digital Innovation – Enel's Experience." Web. October 20, 2014. http://www.engerati.com/article/digital-innovation---enel's-experience.

[7]   Smith, Mike. "The Enterprise Comes to Baltimore." *Utility Analytics Institute.* Web. May 21, 2014. http://www.utilityanalytics.com/resources/insights/enterprise-comes-baltimore.

CHAPTER 1

The Inevitable Emergence of the Smart Grid

Introduction

On March 5, 2004, Andres Carvallo defined smart grid as follows. "The smart grid is the integration of an electric grid, a communications network, software, and hardware to monitor, control, and manage the creation, distribution, storage and consumption of energy. The smart grid of the future will be distributed, it will be interactive, it will be self-healing, and it will communicate with every device."

He also defined an advanced smart grid as follows. "An advanced smart grid enables the seamless integration of utility infrastructure, with buildings, homes, electric vehicles, distributed generation, energy storage, and smart devices to increase grid reliability, energy efficiency, renewable energy use, and customer satisfaction, while reducing capital and operating costs."

The DOE released a handbook on the smart grid in 2009, and in the first few pages, made a distinction between a "smarter grid" and a "smart grid." By this reasoning, the former is achievable with today's technologies, while the latter is more of a vision of what will be achievable as a myriad of technologies come online and as multiple transformations reengineer the current grid. The DOE vision for a smart grid uses these adjectives: intelligent, efficient, accommodating, motivating, opportunistic, quality-focused, resilient, and green.

In effect, all definitions of the smart grid, envision some future state with certain defined qualities. So for purposes of discussion and clarity, we have adopted a convention for this book in which we refer to smart grids today as first-generation smart grids, or SG1, if you will. Our vision for the future we define as second-generation smart grids, or SG2, or as in the title of this book, we simply refer to the advanced smart grid. Finally, at the end of this book, we envision a future where the smart grid has evolved to an even more advanced state, which we call SG3.

We use a key distinguishing feature to mark the difference between smart grids as they're envisioned today and how they will evolve as experience is gained and a more expansive vision—our more expansive vision, we hope—is adopted. The difference, while it may seem trivial at first, is fundamental, and that has to do with the starting point for the smart grid project. If the project starts off with an application, then that smart grid by our definition must be an SG1 project. If, on the other hand, the starting point is a deliberate architecture designed with integrated IP network(s) that supports any application choice, then it is an SG2, or an advanced smart grid project.

Nearly all smart grid projects today start with a compelling application, whether generation automation (e.g., DCSs), substation automation (e.g., supervisory control and data acquisition (SCADA)/energy management system (EMS)), distribution automation (DA) (e.g., distribution management system (DMS), outage management system (OMS), or geospatial information system (GIS)), DR, or meter automation, and then design a dedicated communication network that is capable of supporting the functionality of each stand-alone application. Evolved from the silos of the current utility ecosystem (i.e., generation, transmission, distribution, metering, and retail services), the first-generation smart grid carries with it a significant level of complexity, often perceived as a natural aspect of a smart grid project.

In fact, a considerable amount of the complexity and cost of a first-generation smart grid project derives from its application-layer orientation (Figure 1.1). Starting at layer 7 of the open systems interconnection (OSI) stack [1], the application layer—regardless of the application—requires complex integration projects to enable grid interoperability, from the start of the smart grid project onward into the future. As additional applications and devices are added to the smart grid, whether as part of the original deployment or subsequently and over time, the evolving smart grid must be integrated to ensure system interoperability and sustained grid operations. In short, starting with the application brings greater complexity, which comes at the expense of long-term grid optimization.

The advanced smart grid perspective begins with a basic tenet. At its core, a smart grid transition is about managing and monitoring applications and devices by leveraging information to gain efficiency for short-term and long-term

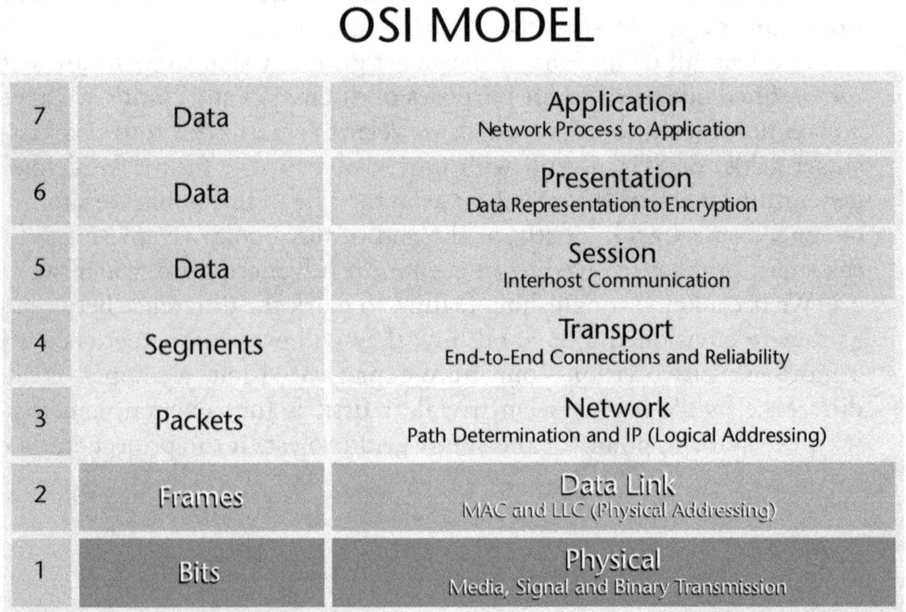

OSI MODEL

7	Data	**Application** Network Process to Application
6	Data	**Presentation** Data Representation to Encryption
5	Data	**Session** Interhost Communication
4	Segments	**Transport** End-to-End Connections and Reliability
3	Packets	**Network** Path Determination and IP (Logical Addressing)
2	Frames	**Data Link** MAC and LLC (Physical Addressing)
1	Bits	**Physical** Media, Signal and Binary Transmission

Figure 1.1   The OSI model or stack.

financial, environmental, and societal benefits. For a system architecture whose principal goal is to leverage information on behalf of customer outcomes, it makes better sense to start with use cases, define necessary processes, choose application requirements, optimize data management and communication designs, and then make infrastructure decisions. A primary focus on the appropriate design process ensures that the system will do what it is meant to do. This key insight—starting at the network layers rather than the application layer—produces the appropriate architecture and design and drives incredible benefits measured not only in hard cost benefits, but also in soft strategic and operational benefits.

Network-layer change stresses investment in a future-proof architecture and network that will be able to accomplish not only the defined goals of the present and near-term future, but also the undefined but likely expansive needs of a dynamic digital future, replete with emerging innovative applications and equipment. A well-informed design and resilient integrated IP network foundation puts the utility in a position of strength that enables it to choose from best-of-breed solutions as they emerge, adapting the network to new purposes and functionality and consistently driving costs out by leveraging information in new ways. The advanced smart grid is foundational; we go so far as to say that its emergence is inevitable.

The advanced smart grid is bound to emerge for two principal reasons. First, electricity is an essential component of modern life, without which we revert to life as it was in the mid nineteenth century. The loss of electrical power, even for just a few hours, is the ultimate disruption to the way we live. We simply cannot live as a modern society without electricity. Second, at its core, technological progress is all about individual empowerment. However, only recently have advances in component miniaturization, computers, software, networking, and device power management technology and the standards that drive their pace of innovation combined to enable individual empowerment in the electric utility industry. A new distributed grid architecture is beginning to emerge that will not only ensure future reliability, but also empower individuals in new ways.

Networks and individual empowerment define twenty-first century technology. It is inevitable that the design of advanced smart grids will begin with a network orientation that is able to accommodate any and all network devices and applications that will emerge in the future. It is also inevitable that advanced smart grids will evolve to ensure an abundant and sustainable supply of electricity and to empower individuals to manage their own production, distribution, and consumption of this essential commodity. The advanced smart grid must be robust, flexible, and adaptable, so it will be; as projects move along the learning curve, society will insist on an advanced smart grid.

The Most Fundamental Infrastructure

We require electricity to power virtually all aspects of our lives today. Electricity is used in the growth, processing, and distribution of the food we eat. Electricity is used to pump and treat the water we drink and use throughout the day. In fact, moving and treating water consumes more electricity than almost any other single function.

Considering food and water—the most basic elements that sustain us— leads one to think of Maslow's hierarchy of needs, a popular theory of human psychology introduced by Abraham Maslow in 1943 in a paper entitled "A Theory of Human Motivation" [2]. Maslow elaborated his pyramid concept more fully in his book *Motivation and Personality* published in 1954 [3]. With respect to Maslow 60 years later, we would suggest that electricity must be added to the base of his pyramid as an essential component of life, as described in Figure 1.2.

Without electricity, our modern life would grind to a halt. Maslow's hierarchy and pyramid have been challenged over the years, but they still stand as a cogent summary of what matters in life, a neat graphic on how we live our lives. By putting electricity at the base of the pyramid, we acknowledge its fundamental nature. Another way to look at the electric grid is as a foundational network that interacts with the other vital infrastructures that support our lifestyles and economy. Consider all other networks that we depend upon that would not operate without electricity; the water system that brings us fresh water, the sewer system that removes and processes our waste, the natural gas pipelines that bring us gas for household purposes like heating and cooking (and to power the base load and peaking power plants that run on natural gas), and the transportation infrastructure with its streetlights, traffic lights, buses, and trains. The telecommunications networks, including wire line telephones, cellular phones, and new smart phones, would not be possible without electricity. The entertainment media we enjoy, from TV to radio to satellite radio, requires electricity. Most recently, the Internet is powered by server farms and lasers shooting light beams down fiber optic lines, where electricity is so critical that massive battery banks back up data centers throughout the system in case of outages. Similarly, our health care networks of hospitals, medical offices,

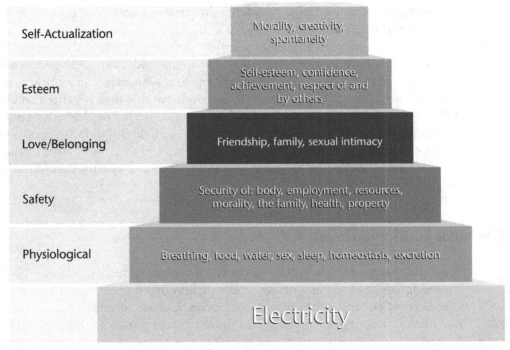

Figure 1.2   Maslow's pyramid, updated.

and pharmacies require electricity as a mission-critical resource, and hospitals in particular rely upon ready access to backup power in a crisis.

We depend on a steady stream of electricity to our manufacturing facilities and power to our homes to run our many household appliances, not the least of which is the humble incandescent lightbulb that started the electricity revolution. The retail stores where we shop use electricity for lighting, heating, cooling, and air conditioning, and to connect themselves to financial networks to process our purchases.

This list truly could go on and on. To drive home the point, electricity and the electric grid have become the mother of all critical infrastructures.

Ironically, the electric grid system here in the developed world has been so stable for so long that we rarely recognize that it's even there, humming away in the background, 35 feet over our heads in a ubiquitous grid of wires, poles, and towers, at least not until we suddenly experience an outage. In an instant, the lights go off, the music stops, the machines lose their spirit, give up their ghost and stop running—and then we count the seconds, minutes, and hours until power is finally restored. Is it any wonder that a blackout is so terrifying when it happens? A blackout is like hitting a giant pause button to stop our modern lives; we must wait for power to be restored before we can go on living [4]. Electricity riots, a foreign concept here in the United States where we may be accused of taking our fantastically reliable electric grid for granted, are very real around the world today [5].

The Drive to Edge Empowerment

Beyond the fact that an advanced smart grid will ensure that electricity continues to stream to millions and millions of power outlets, the advanced smart grid will arrive soon for another key reason. The principal driver of our modern economy today is the unrelenting march of technological progress pushing ever more computing and communications capability, and now energy production technology, out to the individual residential or business consumer on the edges of our networks, supplementing or even supplanting one resource or another formerly at the center of a vital distribution network. Digital technology has let the genie out of the lamp; now millions of scientists, engineers, and business people work around the clock to invent and bring to market a never-ending line of products and services based on incremental advances that empower the edge.

Ever since the integrated circuit came on line in the 1950s, the miniaturization of silicon chips and expansion of capabilities, now widely recognized by Moore's law [6], has proceeded down from micro to the nano level, putting ever more computing power on an ever shrinking piece of real estate. Complementing the march of technological progress in computing power has been a steady advance in communications technology, with fiber optic technology revolutionizing the wired world and wireless advances creating a family of options for sending and receiving radio signals. In the meantime, the Internet has extended itself to every corner of the world in remarkable speed in two decades, going through several phases, morphing into an ever more powerful force for delivering technological empowerment to the edge.

Beyond Moore's law, Metcalfe's law [7] comes into play at the telecommunications and Internet levels, suggesting that the value of a network is proportional to the square of the number of nodes on that network. As more and more devices are added, the network becomes ever more valuable because the number of connections goes up so rapidly. While the exact value proposition of Metcalfe's law has been subsequently challenged in light of the dot-com bust in 2000 [8], the gist is that networking adds tremendous value (we'll keep citing Metcalfe's law as shorthand for the value proposition of networking).

To understand the impact that Moore's law and Metcalfe's law have in value expansion and on potential development milestones for the electric industry, let us compare development milestones in the IT and telecommunications industries to what we project for the electric industry, as shown in Figure 1.3. Mainframes and central switches were not eliminated when distributed options came online, but the focus did shift to the new distributed edge, where greater computing power and communication flexibility enabled laptops and cell phones. Similarly, it is not hard to imagine today's massive central power plants gradually losing their dominance as more and more distributed generation comes online. The reason these trend lines are so similar is that they are all based on fundamental aspects of digital and network maturity: Increasing computation and mobility lead to greater enablement at the edge.

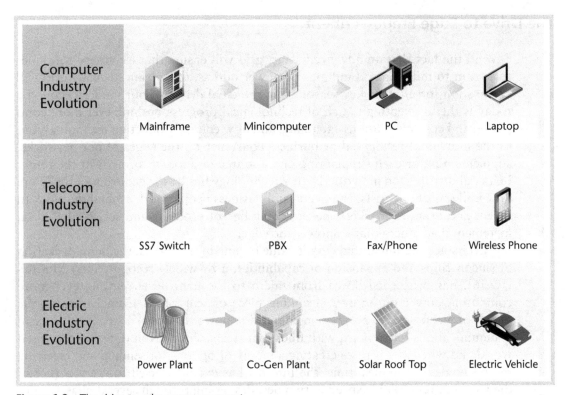

Figure 1.3 The drive to edge empowerment.

The Roots of Smart Grid

Today's smart grid projects have their roots in converging streams of innovation over the past two decades, which, in turn, built upon historic innovation from about the 1950s on (Figure 1.4). Rudimentary supervisory control in the 1950s used pilot wire systems, tonal systems, and twisted pair copper wires. Likewise, early telemetry was employed to provide power readings on remote elements of the grid using multiple communication media, including pilot wire, power line carrier, and microwave systems. The growth in computing capacity and solid-state systems led to SCADA systems in the 1960s.

At the ends of the SCADA systems within utility substations were remote terminal units (RTUs) connected to batteries, which enabled energy control center operators to have visibility and control of multiple elements of the grid within a substation. Inside the control center, map boards that displayed the one-line diagrams of the system were complemented (and eventually replaced) by CRT screens and display software that evolved to provide greater visibility and control. Gradually, as system complexity grew and the grid expanded, humans ceded management functions to increasingly automated systems in order to promote more efficient central control, thanks to emerging technology.

In this way, advances in telecommunications and IT paralleled and complemented advances in power systems. Distribution and substation operations became more and more automated as the system grew in complexity. DA holds greater promise still going forward as the core component element of an advanced smart grid system, where integration is the next logical step in this long evolution for greater control and efficiency.

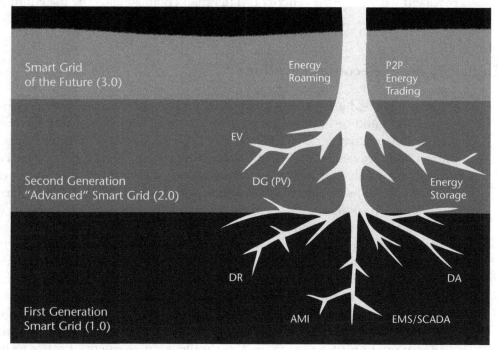

Figure 1.4 Roots of smart grid.

And yet, throughout this long history of evolving complexity and greater use of technology, out at the very ends of the distribution system the analog meter remained, its internal gears driven by electric current flowing through the meter, with corresponding dials under the glass counting off the kilowatt hours, there to be read manually by the roaming meter reader once each month. Solid-state technology employed in multiphase meters came to commercial accounts long before new technology emerged on the residential side, where good enough was sufficient to provide the minimal requirement to produce a bill, a monthly read on kilowatt-hour consumption.

However, by the 1980s, technology came to the residential meter with drive-by automated meter reading (AMR) as offered by Itron [9], where equipment in a trolling van received short-range narrowband radio signals to read new digital meters far more efficiently than a team of walking meter readers could. Similarly, the TWACS (DCSI, now Aclara [10]) and Turtle (Hunt, then Landis+Gyr, now Toshiba) [11] products were early efforts to read meters remotely over narrowband power line carrier (PLC) technology, which became popular in less densely populated rural districts [11]. Early AMR was sold principally as a laborsaving solution, replacing the meter reader but still producing a once-a-month read to generate a monthly electricity bill.

In the mid 1990s, Cellnet (Landis+Gyr/Toshiba) [12] came on the scene offering a revolutionary fixed wireless system (i.e., RF mesh), where meters transmitted to collectors located in contiguous cells that collected data and sent the data in regular bursts to still larger cell collectors for wired transmission back to a central hub. For the first time, interval data was possible to provide a more detailed picture of consumption than a monthly read, but corresponding billing systems would have to catch up in order to take advantage of all that meter data. In time, narrowband RF mesh AMI vendors proliferated, complemented by the continued use of power line carrier systems for more remote service territories.

Electricity and Telecommunications

Utilities have had a back-and-forth relationship with telecom providers over the years. The build-or-buy discussion has proceeded in every utility, as managers weighed the relative merits of investment in communications infrastructure versus outsourcing. Advances in fiber optic technology in the early 1990s proved a boon to utilities, which took advantage of such advances to create new utility telecom divisions and subsidiaries. PLC blossomed into broadband over power line (BPL), and millions of dollars were spent to determine if the power line assets of utilities could be repurposed as communication assets (see Current Group [13]). From proprietary microwave technology, wireless technology progressed in the early years of this decade to open standards–based Wi-Fi, WiMAX, and ZigBee. This book continues the debate over the relative merits of narrowband versus broadband and mesh versus point-to-multipoint network architecture.

Handheld radios, digital pagers, and later cellular phones and wireless laptops kept line crews in touch with central dispatchers. Similarly, technology found its way into utility offices over the years, paralleling the growth of office technology in other sectors. Billing systems, accounting systems, work management systems, telephone systems, desktops, and servers and data storage grew under utility IT department supervision. As utilities moved from mainframes to minicomputers to PC systems, they grew accustomed to connecting with the outside world in new ways, but moving beyond the billing envelope and the telephone as the primary means of communication with their ratepayers proved a challenge to utilities. Gradually, utility Web sites proliferated, and today, social technologies like Twitter and Facebook offer utilities robust tools to interface with their external stakeholders.

If this short review of utility technology shows anything, it shows an already complex industry adapting to increasingly complex demands by incorporating new tools and technologies as they became available. The IT, telecommunications, and electricity sectors grew hand in hand. The electricity from the grid was vital to feed the IT and telecom revolutions. Likewise, IT and telecom advances have proven equally fundamental to the management of an increasingly complex electricity grid. By the first years of the new century, the deep thinkers in this industry who contemplated the long-term future of the grid began to conclude that increasing complexity and continuous technological improvement would lead to ever greater intelligence and automation in the grid.

Defining Smart Grid

The Electric Power Research Institute (EPRI) has promoted the Intelligrid concept for nearly a decade [14]. For almost that long, IBM has been consistent in its effort to draw attention to the need for an intelligent utility network [15]. Rick Nicholson suggested in late 2003 (while at the Meta Group) that the emergence of a new electricity grid architecture would ideally parallel that of the Internet, which he termed an electric geodesic network [16]. Carnegie Mellon, now the home to the Smart Grid Maturity Model [17] initiated by IBM and others, is also a leader in R&D regarding ultralarge-scale systems. At individual utilities like San Diego Gas and Electric, Southern California Edison, and Sacramento Municipal Utility District, experimentation has proceeded apace. At Austin Energy, smart grid innovation expanded to integrate the electric grid with buildings, homes, and transportation infrastructure. Like a family on holiday gathered around a dining room table working on a jigsaw puzzle, leading thinkers have pored over new combinations of technology, sowing the seeds for today's discussion about the smart grid.

Two major themes stand out in this brief review of the roots of today's smart grid evolution. First, vertically integrated utilities are traditionally organized in silos: generation, wholesale operations, long-distance transmission, local distribution, retail billing, and customer service. These distinct business lines depend upon each other but operate with great independence. Consequently, it's not surprising that the vendor community has evolved over more than half a century to meet specific needs in each of these silos. A few very large companies like Siemens, GE, and IBM provide products and services across the silos, but an army of specialized vendors serve single silos as well, meeting the needs of demanding end users with compelling problems to solve. Progress has been measured by innovative individual applications that do more things more efficiently. The idea of a new network approach to solve this ever expanding complexity is novel, but growing, which leads us to the predominant theme of this book.

Design: The Twenty-First-Century Smart Grid Challenge

Today, our grid needs more than a facelift, it needs to be redesigned. Today's electricity grid was designed using nineteenth- and twentieth-century sensibilities, under the limitations of the technology from a bygone era. Few would challenge the tremendous positive impact that the modern electric grid has had on the lives and economies of the residential and business customers who have enjoyed reliable electric power drawn down from the grid. However, the grid—indeed, the entire electric utility industry and business model—is increasingly challenged by the need to provide ever more power at better quality yet reduce costs, by the need to accommodate new and innovative but highly disruptive technologies that in many cases actually reduce revenues, and by the need to reduce a historic, overwhelming reliance on fossil fuels. The only way to address these compelling challenges and conflicting priorities and achieve sustainable long-term success is an overhaul of both the architecture of the current grid and the business model that relies on commodity kilowatt-per-hour delivery. Answering questions such as the following is the reason for this book.

- How did we get here?
- Where do we begin?
- Where should we go?
- Why go through with this?

In contrast, consider today's telecommunications network, which was originally developed about the same time as the electric grid, and under similar technology conditions. While the evolution of the telecom network paralleled the evolution of the grid, telecom development differed by the presence of a huge monopoly. That huge monopoly went through a more radical redesign, first with the breakup of ATT in 1984 [18], then with the incremental addition of wireless services starting in the mid 1980s, the Federal Telecom Act of 1996 [19], and then with the more gradual but more radical emergent Internet.

Like the electric grid, the telecom network was not immune to significant change, which had started much earlier. Midway through the twentieth century, digital innovation from Bell Labs radically transformed the potential of telecommunications, bringing about digital switching, fiber optics, and laser technology to enable ever greater efficiency, capacity, and speed. However, the radical redesign of the network architecture that would ultimately take telecommunications to an altogether new level required three major milestones: first, a decision by the federal government to break the dominant monopoly into pieces to engender competition; second, the birth of cellular wireless to foster a mobile focus; and third, the emergence of the Internet as a fundamental information infrastructure to reshape the modern economy.

Arguably, the electricity grid has not faced such pressure for radical redesign yet, but we're getting there. To draw the contrast, the DOE introductory *Report on Smart Grids* [20] tells a story about Alexander Graham Bell and Thomas Edison, contrasting changes in telecom and electricity. Were Alexander Graham Bell to return today, the story goes, he would not recognize the telecom network elements in quite the same way that Thomas Edison, on a similar journey into the future, would recognize an electric utility's key components. To be sure, the electricity grid has not stood still—it continues to evolve to meet ever greater demand and to address shortcomings and risks, which range from the truly rare massive blackout to the far more frequent and numerous minor outages. However, the grid's historic design constraints, such as a reliance on just-in-time production to meet demand and the need for ever more delicate balancing of grid voltage put the grid at increasing risk of disruption and limit its economic potential as complexity increases and new threats emerge.

In fact, the modern electric grid has long been a model of reliability, especially considering its sheer scale and complexity. Given the critical nature of the grid and the need for reliability, the potential risk and cost of significant change, the historic conservative nature predominant in the utility culture, the lack of leadership at the federal level, and the unprecedented nature of a grid redesign, it is no surprise that any call for such dramatic change would see a slow, even skeptical response. However, the potential for a solution is here today, and this book states unequivocally that these changes must happen and demonstrates the power engineering concepts needed to light the way forward, putting our essential, foundational network back on more solid ground. The electric industry has yet to fully embrace the recent lessons learned from the Internet, which provides not only new ways to operate at the granular level of tools and applications, but also an entirely new architectural model to accommodate the emerging needs of this new century.

Nature and the Internet: Models for Organizing Complexity

The human body is a shining example of the way that networks organize complex systems. Each of the systems inside the human body can be seen as a highly adaptable network, more accurately nested networks, working together under the central control of the brain, but also replete with autonomous behavior apart from central, top-down control. When a person's hand touches a hot stove, for instance, by reflex it draws away immediately based on preprogrammed intelligence that resides in a different part of the brain away from conscious thought, as a matter of survival. No conscious decision is made to withdraw the hand from the heat; it is reflex reaction to the heat. In a similar way, the advanced smart grid will see autonomous behavior from preprogrammed control messages that go out to the edge. The future of the electric grid lies down this path: to mimic the architecture of both the Internet and natural networks and systems to enable sustainability and provide the ultimate in adaptability.

From the outset, the Internet was intentionally designed to be a network solution to provide greater resiliency and to be able to adapt to ever greater complexity. Its elegant architecture has allowed it to grow rapidly and adapt to new applications, new uses, and a dramatic expansion of traffic. In turn, the Internet has taught the world valuable lessons on the way that networks work and provided great insights into how the natural world works as well. Thanks to the Internet, we have also come to understand how the incredible complexity of the natural world is accommodated not just through the trial and error of evolution, but also through the use of self-healing networks that organize and adapt to ever changing individual network elements that combine to form a sustainable ecosystem.

Digitization introduced the grid to the tremendous potential for change in the 1960s, and decades later, the Internet has doubled down on that promise, adding the potential of ubiquitous, high-speed connectivity as broadband technologies gain widespread adoption. Digital technologies have been implemented by electric utilities throughout the supply chain over the years. Both in the corporate offices and out in the field, utility managers worked with vendors to replace their analog devices and solutions with digital devices and applications. However, out in the field, special wireless networks accompanied digitization and soon utility telecom personnel found themselves supporting multiple, incompatible networks. Such a situation was not intentional, but came about because utilities, like most large organizations, organized themselves into independent silos.

So it has happened that utilities adopted Internet architecture in their office environments like any other large organizations, but their field operations ended up evolving down a different path. Applications customized to meet specific needs now prevail out in the field, and more often than not, the supporting networks for these applications are not IP-capable. Vendors have historically offered specialized networks, not only to optimize technological performance, but also to make it hard for competitors to gain traction in their accounts. Understandably, this works out well for vendors, but utilities pay a price when they are left to support multiple networks and are locked in to a specific technology and a specific vendor. This book shows that IP has become so dominant and that security protocols have developed sufficiently such that the time has come to take full advantage of all that IP and Internet architecture have to offer, both in the office and out in the field. The time has come for an integrated, IP networking environment in the electric utility grid, as evidenced by the themes outlined next.

The Inevitable Themes of Change

One may well ask why it is inevitable for the electric grid to evolve into a new architecture that more closely resembles the Internet and the networks in nature, as the title of this chapter proclaims. Is anything truly inevitable? In this case, yes, the transition to a new architecture is inevitable, because the geodesic Web design is superior for flexibility and adaptability in a highly dynamic and unpredictable environment, and the current design, as this book shows, is neither suitable for the evolving nature of the grid nor sustainable over time. It may be argued as to when the transformation will take place, but take place it must. There is too much at stake, and there are no suitable alternatives.

A handful of principal themes work together to drive the discussion on why new grid architecture (i.e., an advanced smart grid) is not only needed, but inevitable. We will show in detail how we can go about getting such a grid. The advanced smart grid is not only inevitable—it is available today. To fully grasp the need for new grid architecture, however, it is vital to understand the significance of the following change themes and what they imply.

Smart Devices and Ubiquitous Connectivity

First, devices and connectivity define the state of the grid—what is connected to the grid and what is not at any given time, and how many connections there are. The emergence of DR and DER, a new class of edge power devices and systems that includes DG, EVs, and ES, means that the industry will need to adapt from connecting a relatively manageable number of devices to connecting a massive number of devices under conditions of far less control. Also, entirely new processes must be incorporated into grid operations. For example, while the relatively few central generation resources are rarely disconnected today (operators ramp them up and down as needed to maintain grid voltage targets), the far more numerous DER devices, including tens of thousands of small generation plants (e.g., rooftop solar PVs and EVs), will in fact need to be connected and disconnected frequently to preserve grid stability. As DER devices gain traction, they will introduce a need for utility managers to promote resource islanding as a strategy to maintain grid stability. (Resource islanding is described in greater detail in Chapter 7.) Moreover, as the number of connected devices increases dramatically, the level of complexity in the grid will rise to the point where automated protocols are needed to maintain stability, and an Internet design will be required to enable the transfer of copious amounts of data and to ensure that the grid remains functional and continues to supply us with the power we are so dependent on.

In the traditional electric utility, a relative handful of generation resources supplied power over the grid to first thousands, millions, tens of millions, hundreds of millions, and then billions of energy-consuming devices. Standards in grid design and standards in operating protocol, as well as standards in the appliances, switches, and plugs on the ends of the network outlined by organizations like the Underwriters Laboratory [21] ensured grid stability and harmony. Even as the number of energy-consuming devices multiplied over the latter half of the twentieth century, maintaining adequate voltage levels was managed by making generators ever larger by adding specialized peaking units and by building more distribution substations to accommodate growth. The grid was able to maintain harmony because the change so far had really been along only one dimension—adding more load—and the solution remained relatively straightforward, if expensive: Add more central generation and beef up grid capacity.

On the other hand, the change driven by technological advances emerging today and the technology that waits on the horizon will be on multiple dimensions. First, the expansion of population and adoption of consumer electronic devices and new appliances means that the number of smart devices and appliances, especially digital appliances, is accelerating the pace of change on the load side. If that were all, it might still be manageable, as more load is merely an acceleration of the trend of load growth we have grown used to. However, a new dimension of change has opened up when DG devices like rooftop solar photovoltaic, microgas turbines (using combined heat and power), and microwind, to name just a few, require power generation control out on the edge as well as back at the core.

DG presents five key differences: (1) utility managers lack the fine control over these new generation devices (assuming that many of them are owned by utility customers), (2) the number of devices is dramatically increased at an unpredictable rate, even though the size of each generator is relatively miniscule, (3) renewable energy generators that depend on the Sun or wind do not produce predictable steady streams of power, but unpredictable variable power, (4) the power is located at the opposite end of the grid, near the load it serves, introducing a new, revolutionary issue—two-way power flow and the need for improved equipment protection—and finally, (5) edge power offers a viable substitute for centralized power as it becomes more prevalent, challenging the exiting business model that depends on growth in load and utility revenues to finance the capital demands of modernizing the grid. The dawning era of advanced smart grids and DER will see millions to tens of millions of new grid-connected devices attached to each grid. In the future, the management challenge will not just be difficult; it will be impossible without significant changes.

Static Versus Dynamic Change

Change—both the pace of change and the approach to change—is a critical management issue in the electric industry. The utility environment has traditionally changed only gradually, and under a controlled setting for the most part, a condition that could historically be described as static. As the transition to a more dynamic state occurs, change is becoming more frequent, less predictable, and increasingly, out of the direct control of utility managers. The utility approach to change management (cultural and organizational) will need to adapt to the new more dynamic state. Consider, for instance, that the current approach to change is based on a specific purpose (e.g., meters, DA, and DR), which retains the silo focus so typical of utilities. In contrast to such purpose-driven change, service-driven change [e.g., service-oriented architecture (SOA)] adopts a more holistic, long view that ensures that all parts of the whole function well together by design, by more fully leveraging a ubiquitous, connected environment.

Consider the pace of change in other industries, specifically the IT industry. Driven by increasing value based on Moore's law and Metcalfe's law, change has become a constant in the IT world, and rapid obsolescence is assumed and built into the product life cycle. As more and more technology creeps into the electric utility industry and as it begins to use more and more IT and telecommunications, is it unusual that the pressure to adapt and change more rapidly would find its way in? When analog meters that had lasted more than 50 years are replaced by digital smart meters, even with strategies that extend product life to 15 years, what is to happen? The utility will be forced to adapt to a new time cycle, where change happens much faster. It will need a new attitude about change, a new approach.

In a static environment with little or infrequent change, it made sense that purpose-driven change would be the norm. In other words, one only changed from the status quo when a specific purpose required change, suggesting a specific application such as DA or DR. However, a more highly dynamic environment, where change is far greater and occurs more frequently (the technological innovation wave reflected by DER alone promises to be highly disruptive for utilities) will require utilities to adopt a new attitude for change, namely, to prepare and acquire the necessary skill sets to make adaptability a core competency. Utilities must adapt to this shift at the organizational level by transitioning away from traditional silos for more interoperability and cross-training; they must adapt at the economic level by shifting their business models to be more oriented toward services and less toward sales of commodity kilowatts-per-hour, and they must adapt at the technology level with a new IP network architecture that will more flexibly accommodate new technologies.

Innovative Design as Change Agent

Innovation at the network level will be led by network redesign as the foundational theme, the adaptation to an exponential increase in the number of connected devices, to new stability mechanisms like resource islanding, to the two-way flow of power and information on the grid, and to a new state of permanent change where everything is in flux. However, control will be maintained by automated policies and protocols out at the edge, by digital devices and intelligence spread throughout the grid all feeding data back to power processors and huge, shared databases, with complete visibility for human controllers there to operate the network with state-of-the-art tools. The new grid architecture itself will have been reinvented from its traditional radial design with relatively predictable, one-way power flow from a few generators out to passive, dumb loads on the edge, to a Web design with highly dynamic, unpredictable two-way power and information flow from hundreds of thousands of generators and storage units sitting alongside intelligent, active loads that also participate in keeping the grid in harmony.

Today, an electric grid design is dominated by two main design types: first, a network design, using N-1 redundancy [22] for reliability flow, that is most often found in more densely populated urban areas; and second, a radial design, which resembles fingers extending out from the palm of the hand, that is more common in suburban or rural populated areas. Current network approaches may be suitable to manage an advanced smart grid, but they might not be affordable. The radial design will certainly not be suitable for two-way power flow. Accordingly, we need a new design, one that is hybrid and affordable and like the Internet, one that will support multiple nodes connected to each other via nested networks with distributed intelligence on every device. We will cover this new design and its merits versus the current available designs in Chapters 2 and 3.

On the supply side, early pioneers started making electricity with small coal-fired generators, but soon the focus shifted to water-driven turbines (hydro) and then back to ever larger coal-fired generators. In time, supply-side innovation added natural gas as a fuel source, then nuclear and further innovation that led to new gas-fired peaking units, first simple cycle, then combined cycle. More recently, renewable energy sources like wind and solar power have emerged to garner our attention. Likewise, demand-side innovation focused first on the incandescent light bulb then went on to a variety of electric industrial and household appliances, from powering commercial ice houses and electrified automobile plants to the electric iron and refrigerator. Innovation on the distribution grid, in between production and consumption, responded to the innovation at the edges, but was driven by silo applications. With an advanced smart grid, innovation by utility managers will be driven by the network. While innovation at the edges will continue, and even increase, the emerging new architecture will bring a renewed, more intense focus on innovation at the network level. In the future, innovative network design will drive innovation at the edges of the network.

New approaches and ways of thinking about systems are already emerging in the smart grid arena. In the distribution automation category, exciting work is under way to integrate three systems that currently operate independently: the GIS, which tracks utility assets out in the field; OMSs, which reactively respond by correlating data inputs during an outage to route those assets to where they are needed; and DMSs, which are being developed to automate many distribution functions and proactively respond to outage information while integrating all the features of a traditional OMS and enabling dynamic GIS evolution. Advanced smart grids require focusing on such key energy concepts as power quality, modulation, harmonics, and fault detection. More information gathered from more places around the grid brings great promise to bring better grid management into the distribution system, improving both routine and exceptional grid management scenarios.

Conclusion

The advanced smart grid is a concept for today, to provide us all with a vision for what lies beyond the considerable work in the trenches now under way in the United States and around the world. While the authors' perspective in this book derives from our direct experiences in the United States, we endeavor to highlight the considerable work and success on smart grid, both in modernizing the grid to accommodate edge power, outside the United States. We provide a framework and a vision that we hope will stimulate a national debate around the important issues that tend to get overlooked when the focus is on solving the urgent, confusing, and complex problems that increasingly dominate the discussion. The smart grid is one of the most vexing challenges humans face today. Upgrading our foundational electric grid while maintaining reliability, overhauling our relationship and our understanding of the nature of energy, cooperating to invent new regulatory, economic, and legal mechanisms and institutions—all these tasks require an inspiration beyond the necessity of short-term objectives, as important as they may be. Herein, as our contribution to the important conversation now under way in earnest on this vital topic, we offer our experience and vision for a future of advanced smart grids that will enable the widespread adoption of edge power devices and systems and ensure a sustainable platform to meet the emerging, dynamic demands of the 21st century.

Endnotes

[1]   The OSI model, often referred to simply as the "OSI stack," is a model developed by the International Standards Organization (ISO) to explain the functionality of a communication system into layers that interoperate in a logical way, by providing unique servers up or down to another layer. The stack starts with Layer 1 and progresses to Layer 7. In Chapter 4, this book presents the fundamental argument that starting a design process after an application choice has been made—starting at Layer 7, in other words—limits the design, whereas starting an architecture design before the choice on applications, avoids such constraints and ensures that the architecture will provide a design that enables the desired functionality.

[2]   "A Theory of Human Motivation," A. H. Maslow, originally published in *Psychological Review,* Vol. 50, 1943, pp. 370–396, http://psychclassics.yorku.ca/Maslow/motivation.htm.

[3]   *Motivation and Personality,* A. H. Maslow (1954), http://www.abraham-maslow.com/m_motivation/Motivation-and-Personality.asp.

[4]   *The City of Ember,* a post-apocalyptic novel by Jeanne DuPrau, made into a movie with Bill Murray, captures the terror of a failing electricity infrastructure; see http://en.wikipedia.org/wiki/The_City_of_Ember.

[5]   http://www.google.com/search?q=electricity+riots&rls=com.microsoft:en-us:IE SearchBox&ie=UTF-8&oe=UTF-8&sourceid=ie7&rlz=1I7ADFA_en.

[6]   Moore's Law, http://www.intel.com/technology/mooreslaw/.

[7]   http://en.wikipedia.org/wiki/Metcalfe's_law.

[8]   http://spectrum.ieee.org/computing/networks/metcalfes-law-is-wrong/0.

[9]   http://news.itron.com/Pages/ami4_0808.aspx.

[10]   http://www.aclara.com/Pages/default.aspx.

[11]   PLC has two definitions that apply to this industry: Programmable Logic Controller and Power Line Carrier.

[12]   http://www.landisgyr.com/na//en/pub/index.cfm.

[13]   http://www.currentgroup.com/.

[14]   http://intelligrid.epri.com/.

[15]   http://www.ibm.com/smarterplanet/us/en/smart_grid/nextsteps/solution/L420447J94627B46.html.

[16]   http://www.idc-ei.com/.

[17]   http://www.sei.cmu.edu/smartgrid/.

[18]   http://en.wikipedia.org/wiki/Bell_System_divestiture.

[19]   http://www.fcc.gov/telecom.html.

[20]   http://www.oe.energy.gov/DocumentsandMedia/DOE_SG_Book_Single_Pages(1).pdf.

[21]   http://www.ul.com/global/eng/pages/.

[22]   http://en.wikipedia.org/wiki/N%2B1_redundancy.

CHAPTER 2

The Rationale for an Advanced Smart Grid

In Chapter 1, we drew a distinction between the conventional view of smart grids and something altogether new, an advanced smart grid, as we introduced a somewhat radical concept. It is inevitable that the advanced smart grid will emerge. The transformation coming to the electric industry as technological innovation crashes over the utility landscape like a giant wave will leave dramatic change in its wake. However, there is more going on here than meets the eye. The role of the utility IT department is transitioning and becoming far more strategic than it has ever been before, as utilities adopt digital devices and shift from the long product life cycles of industrial, electromechanical equipment to the more dynamic, much shorter IT product life cycle, and as digital equipment in the field is networked into an energy ecosystem of interconnected devices, massive amounts of new data will need to be processed, stored, and accessed to and from common servers and databases. The growing importance of data and the shift from vertical silos to a horizontal energy ecosystem will ensure that the IT department in utilities becomes a strategic function.

Introduction

Adapting to these changes will sorely challenge a staid utility organizational culture, but it will also lead utilities to solve challenges that have long bedeviled them, from how to flatten the utility load curve to how to reduce line losses, from how to improve customer service to how to lower the utility carbon footprint. As change progresses, the benefits of change will become ever more apparent, and utilities throughout the industry will become more concerned with how to adapt to change than they will with current challenges that ask if change is even a good idea to begin with.

Understanding the "why" of change lays the foundation for understanding the "how" of coping with the challenges that change brings. So here in this chapter, we dig deeper to explore the rationale for an advanced smart grid. First, we'll describe how such transformative changes lead to the emergence of new energy architecture to meet the needs of a new energy economy. Such changes will inevitably challenge existing business practices and processes and ultimately demand a new set of rules, assumptions, and organizing principles. The changes will take years to accomplish, given the foundational role the grid plays in our lives and our economy, so we will have time to work through the complex details to get to the other side.

We'll explore why a network orientation drives the subsystems within a network to become more integrated with each other and how that must happen. Finally, we'll discuss the planning elements required if such an advanced smart grid vision is ever to be achieved.

A New Set of Rules and Assumptions

The new set of rules, assumptions, and business practices will develop in three principal areas, each building successively on its predecessor: security, standardization, and integration. Security has been a key focus of the National Institute of Standards and Technology (NIST) Security Plan [1], issued in August 2010, as well as the NIST Smart Grid Interoperability Panel (SGIP) [2], spelling out a baseline framework for development of interoperability standards that will guide industry as it creates the new smart grid economy. The reality of a networked economy is that all network components affect each other with varying levels of antagonism, constraint, and synergy. Thus, integration will become more and more critical as it becomes increasingly apparent that the old silo approaches that previously separated parts of the utility from each other for improved functionality now inhibit adoption of new network efficiencies.

Security

The growth of the advanced smart grid depends upon the development of security technology. As the foundational infrastructure, the smart grid cannot afford to get out in front of its ability to remain secure. Economic performance, as well as public and personal health, safety, and welfare, all depend on maintaining a secure and reliable electricity distribution network. The challenge will be to ensure that the grid remains secure as a resilient, mission-critical network capable of providing reliable, affordable power, even as it transforms into a new, very different infrastructure with capacity for so much more.

While the utility industry has seen considerable progress toward ensuring smart grid security, it has also acknowledged the considerable challenges that remain, even as it works to develop a more robust, standards-based approach. Security is unavoidably a dynamic challenge, where systems built to defend against a range of threats must evolve as the threats themselves adapt and change. In a world of dynamic threats, there is no place for a static security solution. NIST first published NISTIR 7628, a comprehensive report sharing a detailed cybersecurity framework and standards, in September 2010 and later revised them in October 2013, to ensure the right methodology and security standards to support the pursuit of secure grid interoperability [3].

NIST leadership on security was welcome news at the time, for without the NIST standards, strategy, and framework efforts, utilities would have found it nearly impossible to hold smart grid technology vendors accountable. Forging security standards is far from easy, however. Electric utilities must invest in three key cybersecurity areas. First, they must invest in preparation and planning to identify the key assets and points-of-contact to secure, design, and update proper assurance and emergency response plans and test and rehearse such response plans. Second, they must invest in mitigation and response activities to monitor events impacting facilities and assets, assess disruption severity, provide situational awareness, and coordinate outage restoration. Third, they must invest in education and outreach to communicate and coordinate with stakeholders, increase public awareness, and form partnerships across agencies, sectors, and jurisdictions. The U.S. Government Accounting Office (GAO) challenged NIST findings on smart grid security in January 2010 [4], and in late January 2011, NIST standards were challenged again when presented to the FERC, followed by an announcement of a three-way collaboration on smart grid security by the U.S. DOE, NIST, and the North American Electric Reliability Corporation (NERC) [5]. The challenge before the industry is to ensure that the promotion and advancement of cybersecurity becomes an integral element of grid transformation. Smart grid solution providers must implement security identity, encryption algorithms, security protocols, and cryptokey management systems that are open and standards-based, robust, proven, scalable, and extensible. Plenty of industry collaboration and guidelines information has emerged since 2010. For example, companies like Cisco, Oracle, IBM, Microsoft, Intel, Alstom, and McAfee have all published detailed reports sharing solutions and recommendations; the National Association of State Energy Officials followed suit in 2011, as did the European Commission Smart Grid Mandate in 2012. Moreover, books on the topic have gone on sale (e.g., *Smart Grid Security* by Gilbert Sorebo and Michael Echols and *Securing The Smart Grid* by Tony Flick and Justin Morehouse). Standards-based security must be designed and deployed into every aspect of the smart grid, supporting governmental and regulatory cybersecurity principles of confidentiality, integrity, availability, identification, authentication, nonrepudiation, access controls, accounting, and auditing.

For the advanced smart grid to be truly secure, security standards must meet the following four minimum requirements:

1. *Granularity at the device level.* Security standards must provide for the identification and isolation of compromised or hacked devices to prevent damage from spreading unchecked through the network.
2. *Standards-based security.* Security standards must be based on best-in-class protocols and requirements that leverage worldwide efforts to develop faster, simpler upgrades to produce sustainable end-to-end security
3. *Multilayer, multilevel security.* Security standards must ensure multiple safeguards in edge devices, embedded applications, network infrastructure, network operating systems, data, and utility enterprise systems
4. *Sustainable security.* Security standards must sustain investment in security oversight, software upgrades, and process improvements and be capable of routine, automatic security updates.

Standardization

We've learned a considerable amount about the value of standards over the past two centuries. In the 1800s, it was common for an artisan or craftsman to work alone and make unique goods. Over time, the business world has slowly embraced the efficiencies and benefits of reaching economies of scale through standards achieved based on a foundation of common protocols, interfaces, and form factors that allow industry stakeholders to evolve their products in a common fashion, differentiating their products by augmenting a standardized product through innovation.

Leaders in the technology era have built on progress in the industrial age using an explicit, strategic focus on standards, gathering together corporate representatives to forge industry agreement on design, production, and operational guidelines that allow products to interoperate efficiently to provide consumer benefits and stimulate demand. Such industry standardization moves the whole industry rapidly through the product adoption cycle and lowers the costs of production as cost efficiencies are achieved.

Standards played a role at the national level in the early days of electricity. Household appliances benefited from industry agreement on common electricity operating designs and UL standards on plugs and switches. Likewise, the PC and consumer electronics industries have benefited from standards, but standards have sometimes lagged the development of new industries, and sometimes they haven't taken hold at all. Who knows where we would be if the world had settled on one global frequency standard—either 50 hertz or 60 hertz—years ago? Still, UL standards managed to make plug and play work for electric appliances inside homes and businesses, if only continent by continent. Groups like IEEE, NIST, SGIP, ANSI, IEC, UCA, and OASIS, on the other hand, seek to drive global interoperability standards for the smart grid that will one day make plug and play ubiquitous on the industrial side of the grid.

However, even as standardization has earned its place in nearly all industries, providing a stimulus for mass adoption and lower costs, the debate continues as proponents of a proprietary approach promote benefits such as innovation and quality control. By building exclusivity into product design and business processes, the proprietary approach creates a buffer between the company and the outside world that allows not just control, but a revenue stream to finance innovation while maintaining quality.

Preventing interoperability is a fair trade-off in this worldview, if innovation and quality control result in superior products and more customer value to drive market adoption. Steve Jobs chose to integrate his Apple software and hardware innovations into a complete user experience with the revolutionary Macintosh, accepting niche status in exchange for what he saw as a superior customer user experience—and he stayed on that path to the end, making the closed loop of a proprietary system a core differentiation with iTunes, the iPod, the iPhone, and the iPad. While it's hard to argue with such proprietary success, the success story of standardization seen in the IBM PC clone shows how an industry standard hardware design and common Microsoft operating system software forged a future that guaranteed ubiquitous computing. Meanwhile, in the digital entertainment and communication sector, a range of alternative products hew to various standards in their existence outside of the dynamic, proprietary "iEconomy." So, this dichotomy, or this dual path between standardization on the one hand and proprietary approaches on the other, coexists in the marketplace.

To date, proprietary devices have certainly been more the norm rather than the exception on the industrial side of electric utility operations. The opportunity of the advanced smart grid is driven by the growing realization that an adherence to proprietary products and processes has become a stumbling block to electric utilities. While proprietary approaches are likely to persist for a long time to come, the electric industry has entered the age of standardization when it comes to the smart grid. Advanced smart grid projects can be expected to take full advantage of the momentum provided by standardization.

Integration

Electric utilities are organized traditionally in silos based on their functional areas (i.e., generation, transmission, distribution, metering, and retail services), and while the silos work together to ensure reliable operations in the electricity supply chain, this structure begins to work against the utility when it comes to the integration of applications and operations over a network. In fact, a principal concern and obstacle to implementing a smart grid is the standard practice of procuring systems that use different communication network technologies (many proprietary and non-IP), that store data in separate databases, and that require separate support systems. Such functional separation, which once provided the benefit of focus and enabled specific functional needs to be addressed, is increasingly counterproductive. Today, a utility that seeks improved interoperability while reducing costs by integrating operations and leveraging a common network and database must replace the efficiency of silo operations with the effectiveness of integration.

With more and more granular decision-making, utility managers need to manage complex and growing databases, and they need access to a universal set of timely data, as well as visibility of system operations of the entire organization. Data can flow to and from a common database to enable individual applications to draw on a comprehensive set of timely data. New operational efficiencies can be identified. An integrated network ecosystem, in contrast to silos, blends the activities inside the utility and among vendors that serve the utility to promote synergy between utility operations, utility communications, application vendors, and network vendors. When viewed from a more holistic perspective, as an integrated ecosystem rather than a collection of partially connected silo organizations, the utility gains tremendous efficiencies. To achieve these and other goals, a focus on the IP network and an accompanying network management system is needed. To create a new energy enterprise architecture and operating system and enable seamless deployment and management of a variety of applications, utilities must begin with a managed network.

Analog-to-Digital Transition

A key driver of the changes under discussion herein is a transition now under way throughout the utility industry, in which analog devices and processes are swapped out for digital versions that can do the same thing, only better and cheaper. Human beings using analog equipment to gather information and make business decisions are being replaced by automated processes using digital equipment and digital communication systems. One industry expert described the changes thus: "Utility control rooms," he said, "once staffed by 50- or 60-year-old industry veterans with

30 years of experience who personally knew all the secrets of the utility—where all the skeletons are buried—those control rooms will in 10 years be staffed by 26-year-olds with graduate degrees, who may have only paltry field experience and little knowledge about the utility service territory, but who will know well how to operate digital distribution management systems, network management systems, and the like, and who will be far more comfortable in front of a computer screen than out in the field, up a pole."

A good part of this transition away from analog instruments and people, in no way unique to the electric utility industry by the way, is based on the maturing Internet, advances in communication systems, and increasing value from Moore's law and Metcalfe's law, which we'll mention again and again throughout this book. Together, these forces drive digitization and its related efficiencies and improved customer value propositions that all work to embed digital technology ever more deeply into our lives. Changes occur not just at the network level, but perhaps more significantly with the upgrading of end devices and sensors and the adaptation of business processes to these new capabilities, architectures, and operational designs.

Consider, for instance, the RTU [6]. The RTU is essential to the operation of a utility substation, and the thousands of dollars for each RTU have long been considered worth the investment. On the horizon, however, are digital routers that will replace RTUs costing only hundreds instead of thousands of dollars—and have greater functionality and versatility to boot. Such change will not just revolutionize the cost structure inside utility distribution networks—as digital transitions like this are implemented, they will free up enormous amounts of capital that then becomes available to finance other digital transitions, creating a virtuous cycle of change. The change, once it begins, will become viral. However, as other distribution industries (e.g., the recording industry and the publishing industry) have learned to their chagrin over the past 15 years, technology-led efficiency can be a blunt instrument that brings with it significant disruption. Such dramatic change must be managed and planned for, so that the medicine meant to cure the patient does not instead kill the patient. What lessons are to be learned from previous significant digital transitions?

One lesson is that a niche of specialized applications will keep the old technology around. The DVD resulted in more perfect reproduction of sound, but audiophiles have held on to their record collections, insisting that DVDs lack the tonal qualities of music recorded and played on vinyl LPs. Similarly, the Kindle [7], iPad [8], and other digital readers have spurred a shift to digital books—Amazon announced in mid 2010 that it sells more digital than hard copy books—but book lovers remain unlikely to give up the textual sensation of turning a page of a beautifully bound book in exchange for being able to have an entire bookshelf in a device that weighs less than a-pound. Analog devices and human processes will remain for a long time in utilities, no doubt, as processes with sustainable value propositions retain adherents. However, the electric utility industry is on the cusp of a tremendous restructuring as each utility reevaluates its core business functionality and asks itself how a digital transition could best meet its specific needs and priorities—to what extent should old equipment and processes be replaced by new ones? Implementation of advanced smart grids will demand such an evaluation; digital devices will replace analog devices only after thorough research and business process improvement (BPI) projects that adapt utility operations to these new realities.

Two Axes: Functional Systems and Network Architecture

To understand the changes technology brings and how the grid will transform, it is helpful to look at two axes that create a change matrix (Figure 2.1). First, the utility business and the emerging smart grid can be segmented along the lines of utility system functions, ranging from central generation and the DCSs that enable efficient generation dispatch to the other end of the spectrum, where emerging DER systems [e.g., solar photovoltaic (PV) systems and EV charging systems] and the smart inverters that will connect them to the smart grid portend a revolution in electricity delivery and consumption. Second, the smart grid system components run from one or more supporting networks on one end to back office servers and databases on the other. The advanced smart grid matrix (Figure 2.1) describes 96 different cost components—four sets of 24. The following section describes the advanced smart grid system that must emerge over the coming decade to support utility operations along these systems and functional areas and with these smart grid system components.

Systems and Functional Areas

DCS

DCS is used to connect the central power plants of a utility with its control center for generation dispatch. This component of a smart grid project involves provisioning high-speed connectivity, generally fiber optics or microwave, between the plants

Advanced Smart Grid Matrix	SYSTEMS AND FUNCTIONAL AREAS							
	Generation	Transmission/ Distribution		Revenue/Billing		Distanced Energy Resources		
	DCS	EMS/ SCADA	DA/SA	AMI	DR	DG	EV	ES
NETWORK								
Spectrum	1	2	3	4	5	6	7	8
Network Equipment	9	10	11	12	13	14	15	16
Backhaul	17	18	19	20	21	22	23	24
END DEVICE								
Hardware	1	2	3	4	5	6	7	8
Software	9	10	11	12	13	14	15	16
Network Operating Software	17	18	19	20	21	22	23	24
BACK OFFICE								
Hardware	1	2	3	4	5	6	7	8
Software	9	10	11	12	13	14	15	16
Network Operating Software	17	18	19	20	21	22	23	24
ANCILLARY SERVICES								
Project Management	1	2	3	4	5	6	7	8
System Integration	9	10	11	12	13	14	15	16
Training	17	18	19	20	21	22	23	24

(Left axis label: SMART GRID SYSTEM COMPONENTS)

Figure 2.1 Advanced smart grid matrix.

and the energy control center and applications that enable interoperability and automated response using separate or shared databases.

EMS/SCADA

EMS/SCADA systems are used to bring back data from distributed elements of the transmission and distribution system for monitoring and control. Subcomponents include RTUs and programmable logic controllers (PLCs). This component of a smart grid project involves provisioning high-speed connectivity, generally fiber optics or microwave, between the endpoints on the network (e.g., the RTUs and PLCs) and the energy control center and applications that enable interoperability and automated response using dedicated or shared databases.

DA

DA includes three principal subsystems: a GIS integrated with an OMS, which will ultimately be replaced in smart grids with a DMS. These systems work together to automate distribution system monitoring and control. This component of a smart grid project involves provisioning high-speed connectivity, generally wireless communications, fiber, or broadband over power line (BPL) communications, between the endpoints on the network—fixed and mobile utility assets—and the energy control center and applications that enable interoperability and automated response using dedicated or shared databases.

AMI

AMI is comprised of smart meter end devices, a wireless communication network, a data backhaul network, a meter head end system, and a meter data management system, integrated to provide interval consumption data collection and processing for use in revenue metering and bill production. AMI systems also provide ancillary functionality, including outage management information, remote turn on/turn off, theft mitigation, and customer data feedback. This component of a smart grid project involves the deployed smart meters, network connectivity (generally wireless, PLC, or BPL communications) between the endpoints on the network and the energy control center, and hardware and applications that enable interoperability and automated response using dedicated or shared databases.

DR

DR systems consist of a remote unit capable of usage data collection, communication, and control, connected to a wireless network and used to automate load curtailment as an alternative to dispatching additional supply resources. This component of a smart grid project involves provisioning telecommunications connectivity, generally wireless or BPL, between the DR device—generally a home energy management system (HEMS), smart thermostat, or a load controller—and the energy control center and hardware and applications that enable interoperability and automated response using separate or shared databases.

DERs

DERs are premise-based systems that produce, store, and/or manage power at the edges of the grid. The principal DER categories include DG, EVs, and ES systems. Each of these DER elements includes some combination of metering and submetering, customer portals, in-home displays (IHDs), building energy management systems (BEMSs), and HEMSs to support functionality at the ends of distribution feeders. This component of a smart grid project involves provisioning high-speed connectivity, generally wired, or wireless, or BPL communications, between each single element or a home gateway and any hardware or applications used to enable interoperability and automated response using dedicated or shared databases.

DG

DG includes any variety of edge-based electricity producing technologies and devices, far more in number, but with far less capacity per unit than traditional power plants. The most popular examples of DG today are natural gas combined heat and power (CHP) microturbine systems and rooftop solar PV systems. CHP systems are popular with industrial and commercial customers that need 24/7 reliable supply. Solar PV systems consisting of solar PV panels and the balance of system (BOS), which may include wiring, switches, racks, inverters, batteries, and/or and net meters. (For pole-mounted systems, land itself may also be included in the BOS.)

EV

EV includes electric or hybrid electric vehicles, electric charging stations, and respective supporting networks. Charging station networks are starting to emerge at residences and businesses, as well as at public locations including curbside chargers and chargers in parking lots and garages. Austin Energy has deployed a large network in downtown Austin, Texas, and charges $5 per month for unlimited charging. Tesla has built charging stations across several states and is planning to have a nationwide network by the end of 2014.

ES

ES is becoming available at multiple levels, including premises, community, and utility-scale. Utility-scale ES is most likely to be positioned in the distribution substation or included as part of a wind or solar farm. Distributed ES, on the other hand, is associated with the load side. Community ES serves multiple collocated residences or businesses. Personal or premises ES is smaller still and generally associated with a single residence or business. A new inroad for ES involves batteries integrated with a solar PV system to enable off-grid operations and avoid demand charges.

Smart Grid System Components

Spectrum and Network Equipment

This category starts with the need to lease licensed spectrum (if unlicensed spectrum is used, there is no need to lease licensed spectrum). The network equipment functional area depends on whether the utility pursues a strategy of dedicated assets and builds its own network or a strategy based on outsourcing—purchase of network services or sharing of carrier network assets. Network functions include (1) last-mile network equipment such as wireless or BPL (this category is the full complement of installed network equipment), and (2) a backhaul system, which provides for the necessary data backhaul between the edge network and the utility core.

End Device

The end-device functional area is comprised of three parts: (1) hardware such as smart meters, which includes the hardware used for specific applications at the edge of the network; (2) application software, which includes the application software needed to achieve end device functionality; and (3) network operating software, which includes the device software needed to manage the device and connect to the utility network operations center (NOC).

Back Office

The back office functional area is comprised of three parts: (1) hardware, which includes the cost of hardware such as servers and data storage devices; (2) software, which includes the utility application software such as SCADA/EMS, databases used to collect and manage data from the head end of the network, and system integration middleware; and (3) a network management system, which includes the software used to manage the network from the utility NOC.

Ancillary Services

The ancillary services functional area is comprised of at least three principal parts: (1) project management, which includes project management to implement a smart grid project (including software, consultant contracts, labor, and overhead); (2) system integration, which includes the process to make the different systems interoperable (including consultant contracts, labor, and overhead); and (3) training, which includes training made necessary by the implementation of a new system.

The New Rule of Integration

A new rule of integration is needed: As we move from application silos to integrated ecosystems, applications procurement by utility divisions must meet network realities.

Spend any time at all around a vertically integrated electric utility, and it rapidly becomes apparent that four major areas of operations distinctly define the traditional organizational culture: generation, transmission, distribution (often shortened to just G, T, & D), and retail operations (i.e., metering, billing, and customer service). These separate functions combine to create the electricity supply chain, and in the slowly changing environment of the past 100 years, this system has worked out quite well for the electric industry and the markets it serves. In certain regions, though, utilities have seen their operations unbundled to facilitate transitions to competitive markets, and the generation and retail operations functional areas at either end of the supply chain have been separated to create altogether unique companies.

When all goes well, the vertically integrated utility is like an orchestra whose parts work together seamlessly to produce beautiful, harmonic music. However, as new functionality is added to utilities (i.e., DER), new areas of operations will raise an important question for utility managers. Should these new and growing areas of operations become new silos that increase complexity further and require significant integration efforts? Or should the addition of such new and different functions challenge the organizational structure of utilities, leading utility managers to consider new options and business models?

In fact, faced with a more dynamic environment over the past several years, and more important, the potential to leverage network technologies and new architectures by working more closely together, this distinct functional separation has become more of an impediment than a boon. We've discussed the issue with silos previously, but now we must dive deeper to understand the true impact of transitioning from vertical silo operations to horizontal integrated operations. The word "silo" may be loosely defined as "a unique area of contained operations that lacks cooperation and coordination with other areas within an organization." While silos will certainly have interlinked operational processes, designed to enable them to accomplish their organizational missions, they often operate independently when it comes to administrative matters, financial measurement (P&Ls), planning, purchasing, and other areas. A common complaint is the duplication of effort that comes from a lack of coordination and poor communication between silos. From a technical perspective, the principal concern is the implementation of systems that require separate support systems, from communication networks to back office IT systems and databases, as described in Figure 2.2.

A significant consequence of organizing in silos is seen in the utility control center, where multiple screens running multiple applications provide controllers separate, distinct views of utility operations, often in different formats, which they must then integrate manually to get a more comprehensive view of the utility system's operations. Unlike telecom operators today, electric utility operators have no view of their entire operations on a single screen. Unlike the telecom operator who can see when a cell phone enters or leaves the network, an electric utility operator has no ability to see single devices; in fact, the operator is blind to grid events beyond the distribution substation. Remarkably, electric utilities still rely upon individual customers to phone in to notify them of an outage in order to determine the extent of a utility outage in any detail out at the edges of the distribution grid.

To examine the silo issue in greater detail, let us consider the application procurement process in a utility department. It is still quite typical for an application to be purchased based on specific departmental (silo) requirements, with little to no coordination with other departments or with the utility IT or telecommunications

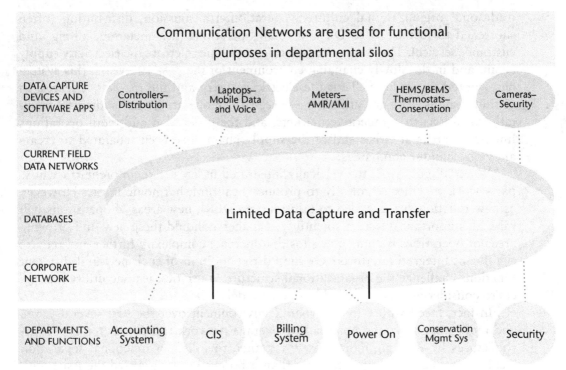

Figure 2.2   Departmental silos and support systems.

department. Savvy application vendors bundle their products into packaged solutions, complete with end-device hardware and software, a proprietary communication network, and a server and database. When multiple departments in a utility follow this process, over time multiple communication networks accumulate; multiple back office systems proliferate; support costs climb as all these networks must be maintained, spare parts must be purchased, people training costs increase, and finally complexity mounts as schemes to achieve interoperability must be designed and redesigned over and over.

With the advent of new network architectures, operations, processes, equipment, and software, an alternative approach has emerged to replace application silos—the integrated network ecosystem. Figure 2.3 describes the potential for a new dynamic process that blends the activities inside the utility and among vendors that serve the utility, highlighting the synergy potential of BPI using four quadrants: on the top, utility operations and utility communications support; and on the bottom, application vendors and network vendors.

Tremendous efficiencies become available with a new set of interoperability standards and processes that stress efficiencies from a holistic perspective (looking at the organization as an integrated ecosystem rather than a collection of partially connected silo organizations). Data can flow into a common database, which enables individual applications to draw on a comprehensive set of timely data rather than more limited subsets that risk leaving blind spots to grid managers. Operational efficiencies can be identified when a complete picture of grid operations is available. Consider that a mere 3% improvement on a $1 billion dollar annual operational budget produces a $30 million dollar reduction in operational

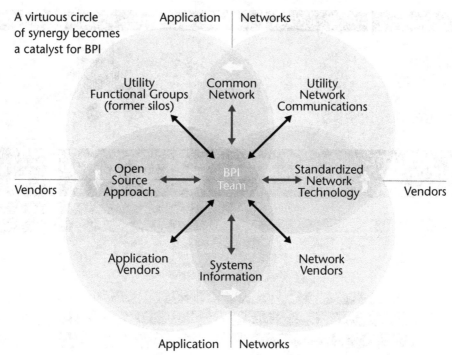

Figure 2.3   Synergistic business process improvements.

expenses—each year—that can be used to defer a rate case, finance other operational efficiencies, enhance shareholder returns, or retire debt.

Integration of Utility Communications Networks and Intelligent Edge Devices

Several disaggregated utility networks operated independently in support of specific applications characterize the current communication ecosystem in a utility. Figure 2.4 details the separate systems that comprise a first-generation smart grid, in support of multiple applications organized in traditional silos. Moving from left to right, note that each system brings with it a separate network to connect the field applications and data with the back office of the utility. Each of these six separate networks supports specific applications within the silos that comprise the different functional areas of a vertically integrated utility, but to function within the utility system, they must be integrated with each other, which requires special software and multiple integration projects.

In this scenario, a system-wide fiber network deployment supports both the DCS that manages the utility power plants and extends out to the substation level to support the EMS/SCADA system and transmission interconnections. DA combines multiple systems, some of which are linked: The GIS is a static asset management system that currently lacks dynamic input into the smart grid; the OMS, a reactive work management system to manage an outage crisis, draws asset data from the GIS and correlates incoming phone calls reporting outage events to work orders for truck rolls; and the DMS, a new system that will ultimately come to

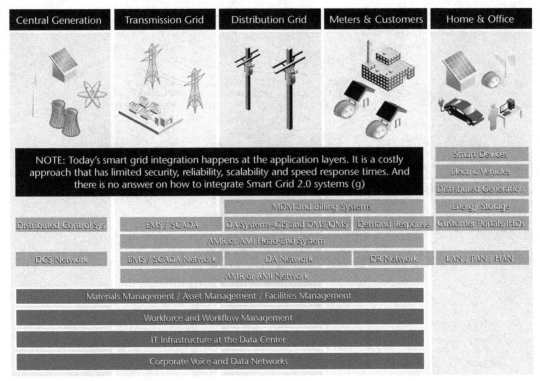

Figure 2.4 First-generation smart grid.

replace the OMS, is comprised of sensors on the distribution grid that communicate through a narrowband wireless network (e.g., 400 or 900 MHz). Likewise, a 900-MHz dedicated wireless network supports the utility's two-way AMI system. The DR system uses a narrowband network that interconnects tens of thousands of smart thermostats with the utility.

While few, if any, utilities have deployed these next systems at any scale yet, the new family of DER technologies depicted in Figure 2.4 is coming online in the next few years and must be integrated as well. This group includes DG (principally solar PV rooftop systems), EV (including charging systems), and ES. Each of these DER systems employs an inverter with a proprietary communication module that will need to communicate with the utility, likely through a shared narrowband wireless network under current operating procedures. This sample firs-generation smart grid deployment, using such a step-by-step incremental approach, could take five or more years to build and require more than 200 separate integration projects.

Power Engineering Concept Brief

Among the many challenges increasingly apparent in deploying a first-generation smart grid, two stand out. First, maintaining a secure network is made even more challenging with an incremental approach. Second, incremental deployment of application-led systems requires numerous complex and costly system integrations projects.

Multiple distinct networks pose yet another challenge. The distinct networks in a first-generation smart grid have unique and proprietary security, different service level agreements (SLAs), different speeds, different coverage, and different costs. In short, they are a complex challenge to deploy, much less to manage. Utilities that purchase their solutions by selecting their applications first, optimizing on the specific functional solution without much regard for network efficiencies and application integration costs, create duplication and add complexity, impose additional limitations, and perhaps most grievously, take unnecessary risks by having so many networks to manage, with neither device interoperability nor end-to-end security.

The operating expense of extra networks, not to mention the additional human resources required to manage and support the networks, is taxing and wasteful, and grows more so as more devices are added. Moreover, given the wide variety of technologies, there are no standard network or performance management tools for most of the networks. Furthermore, most solutions available today still do not offer end-to-end cybersecurity, so that utilities must address not only device security, but also software security, network security, and utility NOC security—all of which are required and must be integrated.

Utilities need more than what first-generation smart grids have offered to date. Yet the current proprietary networking solutions of a first-generation smart grid are incapable of providing essential qualities needed going forward; full quality of service (QoS), virtual private network (VPN), intrusion detection, and firewall capabilities. A true end-to-end smart grid solution must integrate substation automation, DA, DG management, load control, DR, and advanced metering. Such an end-to-end solution must include the strongest security protocols and standards available, including PKMv2, CCM-mode AES key-wrap with 128-bit key, EAP/TLS (with x.509 certificates), and IKE/IPSec, as well as intelligent, standards-based remote smart grid device monitoring and management via a proven, adaptive, smart grid network management platform and network operating system.

Quite simply, system integration remains a huge challenge. To build a smart grid under first-generation smart grid conditions, one must engage in integration at the application layer, which requires the purchase of a middleware solution and interconnection of all the corporate and engineering applications to be able to provide management oversight. With each new application added, the number of system integration projects grows according to the equation $[N^*(N-1)]/2$, where N = number of systems. As N grows larger, the number of projects becomes costly and unmanageable, and the security of the system is challenged. An alternative to this profusion of integration projects is to implement an enterprise service bus, a considerable expense, but worthwhile to avoid the growing system integration expense.

Another costly project element is oversight, which requires the ability to capture key performance indicators (KPIs) and to provide reporting and decision-making dashboards for operational managers and executives in the company. Such integration projects are not trivial—their cost can mount to be as much as twice the cost of the software and the hardware that runs the software. Moreover, the labor commitment remains considerable, even after paying high-priced consultants to complete the project, as staff must be trained to take over, maintain, and run the solutions. Worse yet, so much system integration weakens the security framework. As discussed in detail above, integrated device security is difficult to achieve, as most devices require proprietary retrofitting after the deployment of each of the original networks, making NIST and NERC CIP compliance very difficult, if not impossible.

In contrast to the first-generation smart grid approach, consider Figure 2.5, which depicts the entire integrated ecosystem of a second-generation smart grid. This approach differs from the previous assessment by using an integrated network approach as depicted in dark gray, which describes a SGOE used to manage the advanced smart grid [9].

The second-generation approach begins with a smart grid architecture design, with a focus on the IP network(s) and an accompanying network management system to enable seamless deployment and management of a variety of applications.

The Advanced Smart Grid Approach

A supporting IP network(s) and SGOE are fundamental to the success of an advanced smart grid project. Wired IP technologies available include fiber, BPL, and Ethernet; wireless IP network technologies include 3G, LTE, and WiMAX. Regardless of the technology choice, IP network infrastructure is used to support all the systems in Figure 2.5, including DCS, EMS/SCADA, DA (includes substation automation), AMI, DR, and DER (i.e., DG, EV, and ES).

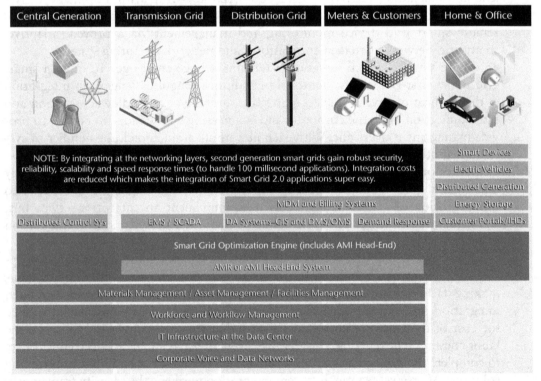

Figure 2.5 Second-generation smart grid.

Power Engineering Concept Brief

The process involved in deploying an advanced smart grid is noteworthy in three key areas. First, the utility pursues access to an IP network capable of supporting all of its communication and application needs going forward. In this build versus buy decision, the utility has three options: to build and own a network or networks, to lease access to bandwidth on a commercial network, or, probably most likely, to choose a hybrid of the two.

Second, the utility leverages a SGOE that enables it to avoid the multiple integration projects required in the first-generation approach above while building the advanced smart grid, but also to do much more. The SGOE provides dynamic balancing of volt/VAR levels based on real-time data inputs from a multitude of devices (where volt is the electric potential difference measurement unit and VAR—volt-ampere reactive—is the reactive power measurement unit for AC electric power systems). However, the SGOE also provides the ability to control the devices and the grid in real time. The SGOE anticipates a much more complex environment, where two-way power flow occurs as the norm rather than the exception.

Third, the utility leverages standards-based digital devices in the field, substituting for proprietary devices that it had previously relied on. For example, digital routers costing hundreds of dollars may eventually replace RTUs costing thousands of dollars. The DCS and EMS/SCADA systems in the second-generation approach will require the connectivity of fiber networks, but the remaining applications will all be supported by a more efficient integrated hybrid broadband network solution.

A New Energy Enterprise Architecture and SGOE

Why are a new architecture and SGOE required for the advanced smart grid? More and more granular decision-making will be required to manage complex and growing databases and use the mountains of data that become available. Utility managers will need an architecture that provides access to a universal set of timely data and visibility of system operations of the entire organization. Accuracy and timeliness will depend not only on knowing which database data is to be drawn from, but also when that database was last refreshed. Without a management system to provide that awareness, utility management will be left with what one utility executive described as "ten thousand versions of the truth." At any particular point in time, a utility manager in an energy control center must have the answer to the question: "What is real, right now?" With inadequate, incomplete, and/or out-of-date information, the definition of reality quickly becomes skewed and highly subjective, leaving managers with a subjective interpretation of reality, and at best lost effectiveness, and worse, risks escalating from there.

If the management vision is to achieve a virtuous cycle of business process improvement derived from a holistic integrated energy ecosystem as described in Figure 2.3, then the current systems of disaggregated utility network communications and disaggregated databases inside a utility in Figures 2.2 and 2.4 must be brought under a SGOE with an integrated, common database schema. While noting that multiple networks may yet persist, it is critical that network traffic is managed from a central point. Only with a SGOE as described above will managers working in an energy control center achieve unitary management vision and control.

Power Engineering Concept Brief

Utilities today have a fragmented view of operations derived from the silos approach and dependence on proprietary technologies that lack the ability to communicate with each other. Beyond operations, the fragmented view impacts utility system planning as well. At the beginning of each week, electric utility managers design an electric network model on paper that is based on anticipated conditions, which describes the current status of all the systems that comprise the distribution grid. However, the planned design they envision is not maintained throughout the week. In fact, a walk through an energy control center today would show multiple operational units monitoring and managing different parts of the grid, from DCS to EMS/SCADA to OMS to AMI to DR, each with a distinct view of the state of the grid provided by the stand-alone proprietary systems. It is left to the human operators in the control center to integrate these disparate views of the grid and make management decisions with the information they have at hand.

The electric network model for any utility is the logical representation of the interconnection of electric elements—including resistors, inductors, capacitors, transmission and distribution lines, transformers, voltage sources, current sources, and demand devices. This model interprets the interaction between the elements of the network based on the rules of physics, including Kirchoff's law [10], Ohm's law [11], Norton's theorem [12], and Thevenin's theorem [13]. The system model anticipates voltage, current, and resistance to help grid operators anticipate system impacts. This method of system planning depends upon a reduction of real complexity down to a defined service territory with a limited number of devices. It monitors the wires and connected devices out to and including the distribution substation. In short, this method relies on a relatively simple model of the complex grid system, projects a static plan, and then changes it based on events throughout the course of the week. This system of artificially simple, relatively manual operations planning is allowed to persist because the number of devices under management still remains relatively limited. Manual system planning may be deemed "good enough" for now, but it will quickly become inadequate with an increasing number of edge devices that will make the concept of two-way load and energy flow a reality.

Features and Benefits of an Integrated Energy Ecosystem

A SGOE, as described above, becomes an essential component of the advanced smart grid. Let us explore in more detail the features of this new visionary tool.

First, a SGOE would need to provide universal management functionality; it should be capable of running on any conceivable IP network (i.e., wired networks such as fiber and Ethernet or wireless networks such as 3G, LTE, and WiMAX).

Second, it is critical that the SGOE provide complete security that is NIST-, NERC CIP-, and FIPS-compliant; it should support end-to-end security, from the devices at one end, on through the software running on the devices, to the network transporting the data, down to the utility NOC presenting the data.

Third, the SGOE should be capable of operating at near real-time speeds—at 100 milliseconds or less—and be able to fully support IP. Instant communication will be needed to support the functionality of an advanced smart grid.

Fourth, a SGOE should provide superb interoperability; it should be able to support all electric devices (e.g., transformers, feeders, switches, capacitor banks, meters, and inverters) from any vendor, because utilities are unlikely to settle for a reduced set of options when it comes to finding the right devices and applications to run their grids.

Fifth, a SGOE should be capable of growing to meet future needs. Such massive scalability will be needed—when the distributed energy resources now under development become commercially viable and begin deployment, millions of new devices will come under utility management purview.

Finally, the SGOE deployed to run the advanced smart grid must not only be affordable, it must also be economically competitive on a total cost of ownership basis: *It must be more affordable than the dedicated multinetwork solution it intends to replace and offer the lowest total cost of ownership (TCO).*

Beyond features, what benefits would be expected to derive from such a SGOE?

First, the SGOE would be expected to provide enhanced energy efficiency, not only improving distribution grid reliability and power quality but also reducing distribution line losses.

Second, the SGOE would certainly provide improved operational efficiency, based on new capabilities such as real-time monitoring and control at the NOC level, self-healing network functionality on the grid, and adaptive distribution feeders to manage the distribution circuits all over the utility service territory.

Third, the SGOE would offer greater customer satisfaction, as proactive outage and restoration services were enacted, as enhanced energy products and services were made available, and as retail energy products were bundled based on targeted customer needs.

Fourth, the SGOE would contribute mightily to soc ietal and utility goals for a gentler utility environmental impact, whether from reduced or sequestered CO_2 emissions, better use of existing infrastructure, closed fossil fuel plants, or from leadership in meeting regulatory requirements.

Finally, the SGOE would provide tremendous economic benefit, as it reduces capital and operating budgets based on its more effective use of system inputs and infrastructure.

A Future of Robust Digital Devices and Networks

A short way to describe the necessary transition to an advanced smart grid could be as follows: When everything becomes smart and networked, the traditional utility becomes an advanced smart grid.

When edge devices become "smart," equipped with IP network communication functionality and localized intelligence, they become capable of being programmed to operate independently and managed remotely, and they can bring back levels of detailed information about the status of the grid never before seen by utility managers and energy consumers. In short, intelligent edge devices suggest a dramatic transformation in grid management capability, as processes designed to improve grid stability by leveraging an abundance of information replace processes designed to maintain grid stability with inadequate information.

In other words, the current grid is designed to operate with inadequate intelligence by using assumptions and estimates to accommodate ignorance, relying on a blend of proactive processes (e.g., tree trimming) and reactive management processes that involve human intuition and intervention (e.g., outage restoration). The advent of connected intelligence portends new approaches that accommodate intelligence rather than ignorance, rely on digital connectivity and automation rather than human intuition and intervention, and proactively diagnose and repair problems to a far greater degree, rather than reactively responding to crises. The advanced smart grid will never be able to eliminate reactive management altogether, given the impact the environment has on the distributed grid. Nevertheless, an advanced smart grid promises to dramatically reduce reliance on reactive processes in favor of far more proactive behavior.

As smart devices are deployed gradually throughout the grid and brought together on a IP network(s), the advanced smart grid becomes the logical and inevitable business model and architecture to replace the traditional utility business model of interconnected, relatively independent departments operating in silos.

Throughout this chapter, we've painted a picture of technology advancing on all fronts into the utility domain. The combination of smart edge devices; a ubiquitous, integrated IP network to connect them all; and a network management system to enable a new set of rules, policies, and procedures describes not just the new advanced smart grid, but also smart networks emerging in other infrastructures that will converge with the grid.

In Chapter 3, we will explore the different infrastructures that support our modern lifestyle and economy and show how digitization and the drive to network intelligent devices naturally leads to smart convergence, bringing separate infrastructures together based on common processes, supporting networks, and databases. What indeed is the advanced smart grid, if not the convergence of the electric grid with the telecom network and the Internet? AMI reforms the revenue collection mechanism of the electric utilities to more closely resemble the relatively mature ATM networks that have provided distributed banking services for decades.

Water and gas distribution systems will also benefit from AMI like electric grids; indeed, these systems have the opportunity to save costs when they share a communications network and related back office systems with an AMI system designed for an electricity grid. Does not the advanced smart grid ensure the changes that are needed for the electricity infrastructure to support the introduction of EVs into a traditionally gasoline-fueled transportation infrastructure? Furthermore, advanced smart grids will lead to the convergence of the electricity grid with our infrastructure of buildings and homes—the built infrastructure—to incorporate both demand response through connected energy management systems and energy efficiency mechanisms.

Endnotes

[1]   NISTIR 7628, *Guidelines for Smart Grid Cybersecurity: Vol. 1, Smart Grid Cybersecurity Strategy, Architecture, and High-Level Requirements,* The Smart Grid Interoperability Panel–Cybersecurity Working Group, August 2010, http://csrc.nist.gov/publications/nistir/ir7628/nistir-7628_vol1.pdf.

[2]   http://www.nist.gov/smartgrid/priority-actions.cfm.

[3]   http://csrc.nist.gov/publications/nistir/ir7628/nistir-7628_vol3.pdf.

[4]   http://www.govinfosecurity.com/articles.php?art_id=3260.

[5]   http://www.oe.energy.gov/DocumentsandMedia/02-1-11_OE_Press_Release_Risk_ Management.pdf.

[6]   http://en.wikipedia.org/wiki/Remote_Terminal_Unit.

[7]   http://www.amazon.com/kindle-store-ebooks-newspapers-blogs/b?ie=UTF8&node= 133141011.

[8]   http://www.apple.com/ipad/.

[9]   The SGOE is described in greater detail in this chapter and in Chapter 7. For now, it is sufficient to consider the SGOE as an advanced network management system for smart grid modeling and operations.

[10]   http://en.wikipedia.org/wiki/Kirchhoff 's circuit_laws.

[11]   http://en.wikipedia.org/wiki/Ohm's_law.

[12]   http://en.wikipedia.org/wiki/Norton's_theorem.

[13]   ttp://en.wikipedia.org/wiki/Th%C3%A9venin's_theorem.

CHAPTER 3

Smart Convergence

In Chapter 2, we explored the rationale for an advanced smart grid, describing the transformative changes that lead new energy architecture to emerge to meet the needs of a new energy economy. We reviewed existing business practices and processes and a new set of rules, assumptions, and organizing principles that arises from the new architecture. Finally, we showed how the presence of a new architecture managed by a smart network enables an array of new smart devices and leads to more and more integration of systems to leverage the new capabilities of the network and its interconnected edge devices. A natural outgrowth of integration within the utility is the convergence of utility functions, driven by two megatrends in the industry and the economy over the past four decades—first, the analog-to-digital transition occurring in devices and processes and, second, the networking of smart devices to drive greater and greater business value, efficiency, and functionality.

Introduction

Infrastructures and technologies are converging around common issues and pressures to become smart, based on these two ties that bind. Digital and network business models, tools, and infrastructures enable electricity (electric utilities), voice and data services (fixed and mobile telecommunication networks and the Internet), entertainment media (cable networks), finance (banking networks), housing (built infrastructure), transportation (e.g., vehicles, roads, and rails), water (water utilities), and natural gas (gas utilities) to become ever smarter in their own ways, each reacting to these two principal business drivers.

The electric industry has a unique position as the support infrastructure to all the other industries and infrastructures, so the changes afoot in electricity will also begin to pull at other industries, long considered to be separate. Still other industries, rather than being drawn directly into a changing utility world, will actively seek alliances with players associated with the electric industry, as it becomes more connected and more automated. The convergence we will see between the electric industry and these other industries will revolve around the ability to connect the dots between networking technologies and principles that derive from our experience with the Internet and the common thread of replacing all analog systems with fully digital systems.

Smart Convergence: Networking Infrastructures, Stakeholders, and Markets

As the Internet has shown us and as we have learned from advances in network science over the past 20 years, the value of networks grows in remarkable ways as more links and more nodes are added. Metcalfe's law highlighted an exponential relationship between network growth and network value; before the network accomplishes any work, the very act of networking, it would seem, adds value. The optimization imperative then becomes managing through the difficulties and challenges of networking to reach the benefits that lie on the far side of the transition. Being connected has another consequence, which is to draw the newly connected nodes in the emerging network closer together and allow far greater levels of intranetwork communication. As electric utilities contemplate networking their systems in new ways, they are drawn to look at their infrastructure and management practices in new ways.

One consequence of networking is to reconsider relationships both within the energy system, between utilities and their power-consuming customers, buildings, and electrical appliances, and outside the energy system, with parallel infrastructures such as transportation networks and water and natural gas distribution networks. The changes contemplated in this book will dramatically impact these other institutions, stakeholder groups, and infrastructures—the smart grid is not the only thing getting smart. We label this movement to smart internetworked systems, stakeholders, and infrastructures *smart convergence,* and we see it happening throughout the new smart infrastructure world.

As smart devices of all kinds proliferate in businesses and homes, each with incredible computing and communications capabilities, the need to monitor and control them is driving electricity managers to borrow from the telecom industry. Furthermore, as the number of smart connections to the grid escalates, a new network management scheme is needed, what we call the advanced smart grid, described in further detail in this chapter. The advanced smart grid makes smart convergence of all types possible; we'll consider just a few examples to start.

First, the role of energy producer and consumer is converging. As energy consumers become energy producers as well, DG (principally rooftop-mounted solar PV systems) will create a need to incorporate onto the grid tens to hundreds to thousands to millions of new power electronics, inverters and net meters, all made smart by adding new computing and communication capabilities. This blending of energy consumption and production represents the realization of Alvin Toffler's paradigm of the "prosumer" in the electric ecosystem (where one becomes both a consumer and a producer). If grid managers are to take advantage of the opportunity to create new resource alternatives that can be dispatched to help ensure grid stability, they will need to find a way to connect and ultimately to control these proliferating edge energy systems.

Second, the built infrastructure is even now evolving to gain new capabilities. Smart buildings and their occupants are beginning to take a more active role in their consumption of energy, thanks to new efficient building design and construction practices and new digital energy management technologies under the collective heading of BEMS, which enable building managers to make more efficient use of the evolving electric grid ecosystem. (For home owners, the equivalent is HEMS).

Third, the transportation infrastructure is evolving to embrace EVs on a grand scale. A smart charging infrastructure in homes and businesses and out in public spaces will be needed to fuel an increasing number of EVs, much as our current network of public gas stations fuels our internal combustion vehicles.

Fourth, ES promises to be a game changer, dramatically altering the capability and business processes of the electric utility, based on the smart charge/discharge functionality of associated energy control systems, as well as the location of the ES units and their interconnection to the advanced smart grid.

Electricity and Telecommunications

Electric and telecom utilities grew up together, kids on the same block. In some ways, these industries have been on a path of convergence from the moment they were born, as we pointed out in Chapter 1's discussion on the electric utility use of telecom technologies. Electricity made possible the invention of telephony, and in turn, the widespread network of power plants, transmission lines, and distribution lines has from the very beginning depended on voice telephony and telegraph networks. In the early twentieth century, these two industries consumed the products and services that the other produced. Later, the electric industry became a principal beneficiary of pioneering advances in telephony and IT at Bell Labs. The telecom industry and Internet pioneers that emerged in the 1990s have built data centers where they found ready access to cheap, high-quality power. For most of their respective histories, however, each industry has regarded the other as separate and distinct. In some ways, each is in the business of distributing a commodity: Telecoms distribute voice connections and later distributed digital bits and bytes, where electric utilities distributed electrical charge for power and light. However, the businesses are sufficiently different that they have remained separate industries serving separate markets.

Yet, over the past two decades, we have seen growing convergence of these two industries. Utilities began investing in fiber optics as an improvement over their narrowband wireless or copper wire infrastructure, the better to support their SCADA systems and other connected devices. Many utility leaders recognized an opportunity in the synergies that come from owning towers, poles, copper wires, and fiber lines and created utility-owned commercial telecom units (UTelcos). A couple hundred utilities now operate UTelcos inside their utilities, engaging in wholesale telecommunications service transactions including acting as a carrier's carrier and leasing dark fiber for long-haul communications, tower and pole mounting rights, and utility rights of way. Some utilities became retail telecom providers after the Federal Telecom Act of 1996 allowed utilities into the competitive local exchange carrier (CLEC) space, but few utilities have had success at retail telephony, and most of those pioneers ultimately pulled back after a handful of retail utility telecom pioneer ventures failed.

On the wireless side, the Utilities Telecommunications Council (UTC) [1] has advocated for favorable radio and spectrum policy for utilities worldwide for over 65 years. Utility telecommunications departments were created to manage the variety of mission critical telecom functions, from radio operations to proprietary wireless networks to the field communications needs of the different departments within the utility, as well as the more conventional telephony needs of utility office workers. Utilities now boast considerable experience as both network operators and consumers of telecom services of all kinds, comprising the second largest market for telecommunications equipment and services. While utilities debate the relative merits today of investing in telecommunications assets and infrastructure (the "build" strategy) and outsourcing their telecom needs to commercial telecommunications companies (the "buy" strategy), a new area of convergence is emerging based on new developments in software and product development.

An area ripe for exploitation by electric utilities is to reevaluate their telecommunications architecture, which has generally been built opportunistically, in ad hoc fashion to support functional applications housed in departmental silos, as described earlier. The opportunity then is to move beyond these silos, designing new smart grid telecommunications architecture from the ground up, in order to model the successful architecture found in telecommunications and Internet networks. In other words, electric utilities must prepare for a future of networking millions of devices under the management of a single utility NOC. In a telecom world connected by ubiquitous IP networks, both fixed and mobile, where intelligent mobile devices proliferate, the potential to reconfigure electric utility telecommunications architecture is breathtaking. Utilities should anticipate an increasingly complex array of intelligent edge devices that will need to be connected and monitored—and ultimately, controlled—as elements of the electricity infrastructure. This first level of convergence is foundational and deterministic to the remainder of convergences discussed below, given that it is the basis of what we call the advanced smart grid.

Power Engineering Concept Brief

Implementing a wired and wireless IP network is the root of a telecom convergence strategic project, but in contemplating such a project, one is led to ask: "If generation capacity is built to meet peak demand, then why isn't a similar approach taken when it comes to customer services and the associated infrastructure to deliver such solutions as DR?" In other words, rather than aligning communication infrastructures and systems to meet peak demand, utilities more often have devised an incremental, lowest-cost approach to telecom to meet their universal service requirement. Approaching telecom infrastructure from a lowest-cost basis and conforming processes and infrastructure to meet minimal requirements, rather than designing telecom architecture to meet peak requirements, is problematic on several fronts.

Consider that to achieve just one application goal—to make DR a dispatchable resource—requires implementing a sufficiently robust communications infrastructure to enable real-time management at the equivalent of peak-driven telecom architecture. However, a conservative culture and economic hard times have led managers to deploy wireless infrastructure that depends on such "affordable, reliable" twentieth-century technology selections as digital paging networks. Given that electricity consumers already live in an on-demand, real-time society and expect services, even utility services, to match current technology capabilities, building a system that relies on twentieth-century functionality, even if it is built at the lowest cost, is bound to come up short in meeting customer expectations.

To address these and other deficiencies and build an advanced smart grid, such logical disconnects must be addressed head on. As an industry, we sometimes forget that our decisions should be rooted not only in lowest cost, but also in QoS and reliability—in short, value should drive decisions. Again, compare the investments to support quality processes and service levels that Japanese grid managers have made over the same time period to those of their American counterparts. Thanks to their foresight and investments in infrastructure and technology, Japanese grid managers now achieve a three-minute annual disruption of power for their entire country—that's a "three-minute system average interruption duration index (SAIDI)" in utility-speak. In contrast, U.S. grid managers only achieve a 120-minute SAIDI. Don't blame them, though—their infrastructure lets them down. It's as if U.S. utilities and regulators, instead of building the right infrastructure to support a best practice use case, have become oriented to an approach based on "safe" incrementalism. In effect, the electrical system built in America lowers functionality standards by default down to the capabilities of the lowest-cost system, instead of raising those standards up to make them world-class, capable of meeting the needs of the new advanced smart grid.

To enable an advanced smart grid, individual utilities will need to undergo a strategic transition away from this ad hoc, cost-justified approach by adopting a planned, robust, sustainable future-proofed network capability and infrastructure. We are, of course, now talking about an IP network with fiber extended to mission-critical facilities to create a foundational supporting infrastructure that is then complemented by a wireless IP network overlay to provide universal coverage throughout the service territory. (For utilities with large service territories that include less densely populated rural areas, different technologies may need to be added to provide cost-effective coverage. For the discussions in this book, however, we remain focused on the distribution grid with sufficient population density to rely on a single wireless IP network coverage solution supported by fiber backbone and backhaul.)

In summary, three alternatives present themselves to achieve wireless IP network coverage: (1) a private, utility-owned IP network; (2) a service contract with a public carrier; and (3) a hybrid private-public network.

Private, Utility-Owned IP Network

The options for a private utility-owned network must first answer the question of licensed versus unlicensed spectrum and the preferred technology solutions most commonly used by each (e.g., spread spectrum and fixed-carrier spectrum, respectively). The advantages of unlicensed spectrum solutions principally revolve around access and lower cost, which are offset by potential disadvantages that include the risks of unpredictable congestion and interference. Unlicensed spectrum solutions have enjoyed a favorable environment at the outset, based on the relative lack of traffic on the spectrum, but as traffic grows from multiple applications, multiple tenants, and a proliferation of devices, the network performance can be expected to decline, presenting a challenge to overcome. For example, a baby monitor sitting on a windowsill next to a meter on the outside wall may disable meter reception that relies on an unlicensed network, if they both operate on the same unlicensed frequency. Users of in-home Wi-Fi experience this conundrum when the microwave oven disrupts their Internet access. In contrast, licensed spectrum solutions offer an exclusive right-of-use advantage but come at a higher cost and require spectrum availability from the holders of the spectrum.

Beyond spectrum and technology used, a private network solution must consider the operations aspect. A network operator, either the utility itself or a contracted provider, will be required to operate and maintain the network. A hidden challenge and cost of a private network solution is technology obsolescence, which presents a risk as new solutions become normative, making spare-part sourcing and maintenance a strategic consideration. Another challenge is that the rapid changes that are common in wireless telecom are out of synch with the long-term nature of utility investment, where expectations of useful life can exceed 15–20 years, in contrast to telecom life cycles of only five to 10 years.

Diving a little deeper, we next consider the strategic network design of a smart grid private wireless IP network, designed from the ground up. The demands of a machine-to-machine (M2M) IP data network are significantly less than the requirements of a commercial wireless IP data network, which must anticipate mobility and high-bandwidth video uses, as well as the need to penetrate exterior walls to reach users on the building interior. Consequently, a network designed primarily to serve a smart grid, lacking such onerous requirements, can have larger cell sizes and still find signal strength at the edge sufficient to maintain connectivity and transmit the data packets as anticipated.

The essence of this innovative approach is to design and build a large-cell "thin" smart grid network that provides signal strength at the edges sufficient to read smart meters. (In Figure 3.1, accepting lower signal strength at 90 dBi enables a coverage area C that is 13 times greater than that of a commercial network design A, using the same network equipment. A dBi or dB isotropic is the forward gain of a wireless antenna compared with a hypothetical isotropic wireless antenna, which uniformly distributes frequency waves and energy in all directions.)

Exponential Leverage:
Coverage Area Grows as Radius Increases

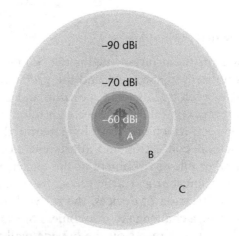

Figure 3.1   Thin smart grid network design.

This breakthrough approach to building a WiMAX network is enabled by creative design parameters, targeted strategies, and adjusted assumptions on such key cost drivers as tower requirements, cell radius, number of cell sites, and spectrum cost. Targeted strategies include infrastructure bartering arrangements with spectrum owners and carriers to leverage utility assets (e.g., towers and ROW) for reduced telecom costs and the "thin" network strategy discussed above, which accepts black spots, to be covered in a subsequent commercial deployment.

When working with a commercial network operator to obtain spectrum rights, a utility can anticipate that the network equipment used in the thin smart grid network may ultimately be incorporated into a denser commercial IP network. Thus, such an initial deployment may be followed by a second phase to deploy additional network equipment to reduce cell size and raise the network capacity to meet commercial standards for providing IP network services.

Public, Carrier-Owned IP Network

The second option—a service contract with a public carrier—brings with it a new set of strategic considerations. The issues and risks of spectrum access, network operations, and technology obsolescence described above now become the purview of the public carrier, which is a big selling point hammered home by the public carriers: "Don't worry, operating, maintaining, and upgrading networks is our critical core competency." However, utilities face a new challenge when considering a public carrier solution: Assuming the cost of access pencils out, utilities must come to grips with the risks associated with placing a mission critical function in the hands of a third party. One utility executive explained in words to this effect: "An SLA with penalties does me no good when power remains disrupted due to a network failure. How can I explain that to a ratepayer?" So, establishing trust with utility executives will be paramount if carriers expect their service options to be embraced by utilities going forward.

Carrier options generally fall in one of two categories, depending on the network being leveraged. Carriers with maturing third generation (3G) networks, whose costs have been borne by voice and data services in the commercial space, now look at these mature networks as logical transport providers for the emerging M2M space, which have light data requirements, but which often require the kind of coverage only a carrier can provide. Since the capital costs of the 3G networks are mostly paid off, services operating on these networks can be aggressively priced based on marginal costs.

Other carriers will seek to take their emerging long-term evolution (LTE) network technology to the marketplace as a platform for both voice and data services, including M2M. A principal challenge for telecoms with LTE will be the immaturity of their networks and the lack of coverage, until such time that their networks are fully deployed. Prices will be relatively high at the outset, but the coverage of the new networks will improve steadily over the coming years. The numerous technical advantages of the new LTE technology are expected to make these network options appealing and to help telecom carriers to win over new converts.

Hybrid Public-Private Network

With telecom carriers aggressively marketing their mature 3G networks in the M2M space, utilities may take advantage of the coverage benefits and relative low costs of a ubiquitous 3G network to provide coverage for meter reading and other device access in hard-to-reach rural areas. Some may consider 3G as a transport solution for meter data in more dense urban areas when costs and internal capabilities make the carrier solution more attractive. However, for mission-critical functions (or urban noncritical tasks), the utility may still choose to rely on a privately owned network that provides greater certainty of performance and public reassurance of reliability—if at a greater cost. Wild cards in this area include aggressive pricing by the carriers, given the relative lack of capital for financing new network construction and the unwillingness of utility regulators to add costs of new networks to the rate base, where an existing carrier network can be used as an alternative at a reasonable cost comparison.

Electricity and IT

The second great convergence, closely related to the telecom/electricity convergence described above, relates to the steady adoption and integration of IT devices, applications, processes, and market concepts into the highly controlled and regulated world of the electric utility, which is charged with "keeping the lights on." Electric system reliability has to be one of the most challenging organizational dictates imaginable. High voltages are deadly; electric lines are mostly above ground and constantly subject to environmental disruption; and the highly fragmented, complex, interwoven electricity grid of today, built on a foundation of aging equipment, continues to grow in size, complexity, and importance as the foundational infrastructure of this digital, networked era. System reliability must address these constraints with the exacting operational standards needed for the grid to perform its functions.

The challenge of this smart convergence is to weed out the "good" digital adoption from the "bad." In a conservative environment, where utility infrastructure investments are subject to public scrutiny by regulatory bodies, local governing boards, and member-representative boards, replacement of working and reliable, if aging, analog equipment with new digital equipment is bound to proceed slowly. Many in the utility industry hold to the "better the devil I know than the one I don't" school of thought, especially in the IT world, where IT vendors have offered an ongoing array of options that promised enticing benefits when gaining approval, but sometimes failed to deliver on those promises after they were deployed.

Moving beyond the reluctance to replace a workable solution with a new, better, or cheaper solution that may or may not work as promised, an additional challenge is a culture that has built a reliable infrastructure using industrial-age business practices, where electromechanical equipment has typically enjoyed a useable life measured in decades rather than years. Beyond the relatively long useful life of electric utility assets is the business practice of leaving working assets in place when they operate well beyond the expiration of their useful lives as recorded in depreciation schedules. This culture has become a principal challenge to the digital transition in practical terms. The conventional wisdom embodied in the phrase, "If it isn't broke, don't fix it," leads to a culture of keeping assets around well beyond their documented useful life, essentially leading to a prevalent business practice of running equipment until it fails.

In contrast, the IT equipment replacement cycle is measured more in tens of months than tens of years. Driven by the economics of Moore's law, IT equipment is routinely obsoleted in a matter of three to four years based on the availability of new equipment that has greater computing power, greater storage capacity, and improved functionality, often at a lower price—in essence, making equipment technically or economically obsolete well before it becomes functionally obsolete. In the face of this culture gap in purchasing philosophy, IT departments must work closely with their internal clients in the utility to explore the relative merits of the purchasing cycle and a new mentality that considers opportunity costs on an equal footing with such things as spending caps, tight budgets, and traditional utility business practices.

Analog-to-digital transition drives the smart convergence of IT and electricity. An electric industry that once defined electricity as high technology at its inception has moved into a new era where the convergence of information via digital technology and electricity will create a new, transformed industry over time. When this transition has reached its potential in decades to come, and electricity has finally become a blending of electricity and information, will we still think of it in the same way? An industry built on estimates, best guesses, and human intuition in times of crisis is moving to a far more exacting industry that relies on facts, data, and information to drive automated processes under the oversight of human operators.

The key example of this type of smart convergence involves the replacement of older analog devices with new smart digital routers at critical points throughout the electric infrastructure. These sensors and control devices become new intelligent nodes on a network modeled after the interconnected Web network architecture of the Internet but also overlaid on the traditional radial network architecture of the electric industry. Anyone who has experienced the intermodal transportation network of a large city understands this type of blended topology: the radial network of urban railways, subways, and trolleys radiating from a city center interconnected with the bus lines and roving taxis traversing city streets that more closely resembles an interconnected web. Operating together, these two networks create a synergy that is better than either could achieve on its own.

The advanced smart grid depends not only on an IP network infrastructure and sophisticated network management tools borrowed from the telecom world, but also on digital intelligence built into smart devices. Smart meters, smart routers, smart inverters, and smart consumer devices, each acting independently on the edge based on programmable logic and algorithms, interconnect with communication networks so they may be controlled remotely as needed from the central utility NOC.

The convergence of IT with electricity will lead to yet another analog-digital transition. The traditional business practice involves human decision-making to keep the grid operational, based on deep human experience, skills, and knowledge and a healthy dose of human intuition. The transitioned business practice involves intelligent devices programmed with the knowledge and skills of the best practices of experienced utility workers, then executed with a precision that those human workers would never be capable of matching.

In digitally converged electric industry operations, there will be fewer workers than in the past, but the skills, knowledge, and experience of utility workers will start with the traditional electricity skills of the electrical engineer, lineman, and energy control center operator and add more IT skill sets from programming to network diagnostics. New energy service jobs will arise for displaced utility workers. Business processes will need to be adjusted as well to accommodate this shift in digital capability and the new labor demands of electric utility organizations. Utilities, with help from government grants and loans, will be in the business of retraining their workforces to operate grids using digital equipment.

Power Engineering Concept Brief

The installation of a ubiquitous wired and wireless network throughout the utility service territory opens the door to an examination of utility operations and the devices used to execute utility functions, for this is the tableau on which the analog-digital transition will be played out in the utility landscape. From a power engineering perspective, the digital applications with potential in the advanced smart grid paradigm include those listed in Figure 3.2, with special attention paid to the key applications described in detail in the following sections.

Condition-Based, Predictive Equipment Maintenance

As stated earlier, the prevailing business practice in most functional utility silos is to react, replacing devices at the point of failure. In contrast, predictive equipment maintenance involves searching for patterns to identify faults inside circuits. Predictive maintenance of distribution assets involves monitoring devices to detect changes in power consumption, which act as red flags. Older devices near failure consume power differently, creating distinctive patterns and signals that can be diagnosed and detected to identify problems before they happen. What's more, anticipated performance based on published specifications can be measured and compared to measured performance in real-world scenarios, allowing for fine-tuning and advanced diagnostics.

DR Management and Analytics

Currently, DR is managed primarily during peak seasons and at peak hours in order to lower system peaks and to avoid uneconomic operations, using technology that in some cases is neither real-time nor two-way, but which still offers a valuable service to help manage the grid better during these difficult periods. Aggregation of a fleet of smart thermostats or HEMSs, however, promises to provide the utility a powerful new resource, only now being realized. Imagine a digitally connected distributed resource of aggregated smart devices as yet another grid management tool, working collectively to fine-tune and optimize grid voltage and current levels minute-by-minute according to preset algorithms that balance power consumer-prescribed conditions with the utility conditions needed to harmonize the grid.

Advanced Smart Grid Digital Applications

- Condition-Based, Predictive Equipment Maintenance/Asset Management
- Demand Response Analysis
- Direct Customer Load Control
- Dispatcher Training and Simulation
- Distribution System Real-Time Analysis Tools
- Emergency System Restoration Support
- Fault Detection
- Feeder Equipment Monitoring
- Integrated Volt VAR Control
- Integration of DMS(OMS)/AMI/GIS
- Load Forecasting
- Multilevel Feeder Configuration
- Network Switching Management/Analysis/Optimization
- Power Quality Assessments
- Relay Protection Coordination

Figure 3.2 Smart grid digital applications.

To get a bead on the potential, consider that today nearly every household has a number of charging devices, which are miniature transformers that often stay plugged in, continuing to draw power after the device has reached full charge, a condition that has come to be called "vampire power." While consumers can be urged to unplug those devices, or to buy special plug inserts to unplug them automatically, many can't be bothered with doing so. Imagine though if this waste of power could be converted in an advanced smart grid environment into a latent asset, lying fallow in wait for a command from grid managers to switch off to help them fine-tune the grid when it is under stress.

Fault Detection, Isolation, and Restoration (FDIR)

As described in Chapter 2, engineers plan the state of the grid using models, planned algorithms, and historical behavior patterns, but these are blunt instruments and the plans thus created soon go stale, since the grid actually works in real time. For instance, when a transformer located out near the end of a distribution feeder has an arcing event that puts it out of commission, the amount of load lost on the feeder line may be detected, but the cause of the loss remains imperceptible from the perspective of a control center operator, who can measure the effect but remains ignorant of the cause. The outage will remain undiagnosed until consumers notify the utility and a truck is dispatched to locate the fault and restore the line. Similarly, when a capacitor bank cracks and becomes the electricity equivalent of a leaky pipe, it ceases to function as planned, and the grid slowly grows more out of balance with low-grade degradation over time. Signal processing, the twenty-first-century cure, correlates and transmits environmental and utility functional data to allow timely comparative analysis of observed and anticipated data and an automated response to put the grid back together again.

FDIR helps achieve dramatic reductions in grid interruptions, integrates reclosers and substation equipment, and ensures better safety, but a critical challenge in restoring power—synchronization—remains. The system average interruption duration index (SAIDI), the system average interruption frequency index (SAIFI), and the customer average interruption duration index (CAIDI) are measures commonly used to report electric service quality. SAIFI measures how often a customer can expect to experience an outage, SAIDI measures how long the customer can expect to wait for power to be restored (regardless of how often the system goes down), and CAIDI measures the average outage duration if there is an outage, or average restoration time. These indices are defined over a fixed time period, usually a month or a year, and can be measured over the entire electric distribution system or over smaller portions of the system, such as an operating area or individual circuit.

CAIDI is perhaps the least straightforward of the indices, but from a customer experience perspective, CAIDI is the most relevant index. While the first two indices are driven by frequency (SAIFI) or time duration (SAIDI), both variables drive CAIDI. Strategies for improving SAIFI and SAIDI can sometimes adversely affect CAIDI. SAIFI is improved by reducing the frequency of outages (e.g., tree trimming and equipment maintenance programs) and by isolating the disruption to reduce the number of customers interrupted when outages do occur (e.g., by adding reclosers and fuses). Strategies that reduce SAIFI also impact SAIDI—an avoided outage has no chance of increasing the duration number. Both SAIDI and CAIDI benefit from faster customer restoration. Perversely, system improvements can make CAIDI go up as well as down, depending on the relative impact of improvements on outage frequency (customer interruptions) and outage duration (customer minutes of interruption). Thus, it's complicated, and interpreting indices requires more than a cursory glance. In the final analysis, all three indices are valuable management tools from an advanced smart grid perspective, because they represent data leveraging to enable grid managers to benchmark and make system improvements over time (Figure 3.3).

System improvement really matters today, given that modern customers have come to expect a higher level of power quality from their electric utility, as they do of the other service vendors in their lives. Digitization in part drives these perceptions as well. Higher power quality requirements have become even more critical with the proliferation of electronic devices such as flat panel TVs, DVRs, VCRs, game consoles, computers, and clock radios, which are intolerant of even the smallest interruption of power and can set lights to blinking throughout the house—just imagine adding solar PVs and EVs into this edge power equation.

Measuring Electric Service Quality

Three key indices measure service quality in the electric utility infrastructure:

SAIDI	SAIFI	CAIDI
(System Average Interruption Duration Index) is the calculation of how long the system remains down after outages during a specified period of time, usually a year, that is, the sum of outage Customer Minutes of Interruption (CMI) divided by the total number of customers served.	(System Average Interruption Frequency Index) is the calculation of how often the system goes down over a specified period of time, usually a year, that is, the sum of outage Customer Interruptions (CI) divided by the total number of customers served.	(Customer Average Interruption Duration Index) calculates the impact of outages on a single customer, that is, the product of dividing the duration (SAIDI) by the frequency (SAIFI).

Figure 3.3   Electric service quality.

One strategy grid managers employ to limit the number of customers affected by an interruption due to a fault is to divide distribution feeders into sections then isolate them using motorized switches or breakers. A new trend is to include smart meters and distribution routers downstream in the feeder for the collection of dynamic data. In this manner, applied FDIR algorithms can detect in which section of the feeder the fault occurred and rapidly isolate that feeder section by operating the isolating switches or breakers and restoring power to the nonfaulted sections, while ensuring that only those customers on the faulted section are affected by the power outage.

Integrated Volt/VAR Control

Put simply, the purpose of the electric distribution network is to move electricity out from substations, then down distribution feeder lines and on to consumers. The distribution system includes medium-voltage (less than 50 kV) power lines, substation transformers, pole- or pad-mounted transformers, low-voltage distribution wiring, and electric meters. The distribution system of an electric utility is a complex system that may have hundreds of substations and hundreds of thousands of components. Most of the energy loss occurring on the distribution system is caused by resistance: Electric current flowing a distance through conductors results in a loss of measured ohms. The amount of the loss is proportional to the resistance and the square of the magnitude of the current. Thus, operators reduce losses by reducing the resistance or the current's magnitude. The resistance of a conductor is determined by the resistivity of the material used to make it, by its cross-sectional area, and by its length, none of which can be changed easily in existing distribution networks. Fortunately, reducing the current magnitude can be accomplished more readily, by eliminating unnecessary current flows in the distribution network.

When evaluating line loss, there is also the issue of active and reactive power to consider. For any conductor in a distribution network, the current flowing through it can be decomposed into two types, active and reactive. Reactive power, measured in VARs (i.e., Volt-Ampere reactive), present in the line contributes to power loss by using up a portion of the current carrying capacity of the distribution lines and equipment. Reactive power compensation devices are designed to reduce or eliminate the unproductive component of the current, thereby reducing current magnitude and, thus, energy losses.

Depending on the types and mixture of loads in the system, the voltage profile on the feeders can also affect the current distribution (and loss of power), although the loss is smaller and its impact less direct.

Traditional volt/VAR control extended along the distribution feeder out to the edges of the grid has achieved reasonable improvement in the distribution grid's operational performance. Voltage regulating devices are usually installed at the substation and on the feeders. Substation transformers can feature tap changers, devices that adjust the feeder voltage at the substation, depending on the loading condition of the feeders. Special transformers equipped with tap changers called voltage regulators are also installed at various locations on the feeders, providing fine-tuning capability for voltage at specific points on the feeders. Reactive compensation devices (i.e., capacitor banks or more informally, cap banks) are used to reduce the reactive power flows throughout the distribution network. Cap banks may be located in the substation or on the feeders and can be fixed or switched.

Traditionally, the voltage and VAR control devices are regulated in accordance with locally available measurements of voltage or current. On a feeder with multiple voltage regulation and VAR compensation devices, each device is controlled independently regardless of the resulting consequences of actions taken by other control devices. This practice often results in control actions that may be sensible at the local level, but contribute to suboptimal effects on a broader scale. More ideally, information would be shared among all voltage, and VAR control devices and control strategies would be comprehensively evaluated to make the consequences of possible actions consistent with optimized control objectives. We term this new approach to distribution system management integrated volt/VAR control.

Such a smart distributed control approach is enabled by multiple trends, including the increasing adoption of substation automation (SA) and distribution feeder automation (DFA), the widespread deployment of AMI, the growing deployments of solar PV systems, and the advent of EVs. All these devices provide the necessary local intelligence at the sensor and actuator levels, based on reliable two-way communications between the field and the distribution system control center, to make distributed control possible. With enhanced sensory data feedback from sensors located further out on the edges of the grid and along the distribution feeder, grid managers are equipped with the local intelligence and two-way communication capability they need to exercise finer control over a larger area of the grid. These new sensors will become even more useful when they are used to help grid operators address the new challenge of reverse power flows coming from such new DER elements as solar PV systems and ES devices located along the distribution feeders.

In fact, the integrated volt/VAR control transforms new DER devices from potential threats to grid stability into new, valuable grid management resources. The advent of the advanced smart grid enables utilities to manage VARs proactively through capacitor banks further out on the distribution feeder, beyond the substation, whether at the pad-mount or pole-top transformer or at the smart meter on the ends of the line.

Integrated volt/VAR control will minimize power losses or demand without causing voltage/current violations, a term that refers to the undesirable excursion from the normal operating current and voltage range for the distribution system (e.g., current that exceeds the maximum safe limit for a given conductor type, voltage that exceeds a limit considered safe for consumers, or voltage that falls short of a limit needed for normal operation). Integrated volt/VAR control solutions would have emerged sooner, if not for the lack of computational resources near the edge, which are needed to solve complex mixed-integer nonlinear and nonconvex problems in order to evaluate the loss and demand for a single functional equation. Solving such equations is at the root of maintaining balance in such a complex system. Such efficiency is critical to address complexity: An algorithm that requires fewer functional evaluations to find the optimal solution will be regarded as more efficient than one that requires more functional evaluations to achieve the same objective.

Electricity and Banking: Smart Meters (AMI)

The revenue meter that hangs on the outside wall (mostly), or the one that is hidden down in the basement or in a closet (more obscurely) is the utility's cash register. Wherever the revenue meter sits, it is traditionally considered the last device on the end of the utility distribution network, and its primary role is to measure energy consumption in kilowatt hours. (With commercial and industrial accounts, it also measures power demand in kilowatts.) The more sophisticated commercial and industrial meters also measure other qualities of the electricity that flows from the grid into the customer premises. Throughout its history, the utility has managed a relatively simple analog system to take monthly readings and calculate energy use based on the difference between two readings at the beginning and end of a billing period. With AMI, remote revenue management using a network is similar to the use of ATMs by the banking industry, providing us with an informative discussion on technology convergence that contrasts the mature network of ATMs with the dramatic expansion of networked revenue meters on the distribution grid.

What could ATMs have to do with AMI? They both automate the delicate task of revenue collection over a distributed network. The ATM network got its start in 1973, when a company called Docutel [2] was awarded a patent for its networked ATM. The trend spread and the machines grew more sophisticated and numerous. Today these secure cash dispensing facilities dot the landscape, and we think little of the revolution they represent. ATMs managed to take the functions of the bank vault and the human teller sitting behind a barred window, which had earlier moved from the bank lobby out to the drive-up window, and distribute them out to the edges of a large network, automating the teller function, but also providing bank customers secure network access to their bank accounts. As such, this transition offers a model of both digital transition and transition to a secure network that has evolved and improved over time, but certainly well ahead of the Internet and the proliferation of Internet security risks and strategies. To round out the picture of networked banking, banks have moved to an Internet model of account access and distribution of bank transaction statements in order to reduce operating expenses and provide more convenient customer access to banking information.

The emergence and growth in the adoption of wireless networked revenue meters by electric utilities offers similar potential to transition electricity consumption monitoring and consumer access away from a twentieth-century model of manual reading of analog meters once a month to produce a paper bill to be mailed to customers for manual payment.

The change potential in automating revenue meter data collection with an AMI is dramatic. Imagine moving from a single monthly meter read (12 reads each year to produce a monthly bill) to a meter read every 15 minutes to produce a bill, but to do so much more beyond managing revenue: Four reads per hour—15-minute interval reads—produces 96 reads each day. That's eight times as many reads in a single day as was produced in a year with the manual, analog system. In a year, the number of meter reads will go from 12 to 35,040. And that's just for a single account. For a utility with one million residential meters, the number of interval meter reads each year by that single utility will be over 35 billion. If each single meter produces 400 MB of data per year, then an electric utility with 1 million residential meters will have the challenge of managing 400 terabytes of new data each year—and that's just for one midsized utility.

What will become of that entire digital interval meter data? Besides being used as a resource to produce monthly consumer utility bills, the massive amount of universal, detailed consumption data opens up a whole world of management possibilities. Consumers will be able to use the interval meter data in HEMSs, whether it is communicated directly from the revenue meter via a technology like HomePlug or ZigBee into the home to an IHD, or communicated via the utility over the Internet and presented on a Web site and accessed via a PC or a smart mobile phone. Utilities will be able to aggregate the data to produce valuable information on system operations and consumer behavior, from a group perspective, and learn new things about how to operate their utilities more efficiently.

For this vision to come to pass, however, the deployed AMI system will need to be physically safe and cyber-secure. The implementation of an AMI system requires not only significant planning, but also ample time during the deployment to test and calibrate to meet the demands of security. Regulations and data privacy standards have made utilities the steward of customer data, and they are obliged to maintain the meter data with care. When it comes to security, the AMI system has a model in the ATM networks that have ensured billions of safe transactions daily.

Power Engineering Concept Brief

As stated earlier, ATM networks, which connect ubiquitous ATMs and are often the target of malicious hackers, provide a useful analogy to the network approach needed for the AMI system that is part of an advanced smart grid. ATM networks have managed to remain well-protected and reliable despite such persistent threats because of the highly sophisticated, standards-based, device-level security that resides in each ATM, which renders it inoperable upon threat detection before any connection can be made to the ATM network, thereby avoiding virus or worm proliferation. While smart device hacks are inevitable, utilities can protect their advanced smart grid from a massive network virus or worm by implementing ubiquitous security architectures. Aligning the architecture on security begins with embedding unique, standards-based hardware and software security into every network device to prevent penetration attacks (i.e., worms and viruses) from spreading throughout the advanced smart grid network. Granular, device-level security in the ATM network quickly identifies and isolates a hacked or compromised device, limiting damage. Similarly, sophisticated, device-level security incorporated in the design of embedded communications devices ensures the most robust protection.

From a technological and engineering standpoint, it should be noted that AMI is not by any means the end of the story. The road from AMR in the 1990s to AMI in the first two decades of our new century has been about increasing real-time distribution management, isolated in the meter as an end device. In an advanced smart grid, AMI undergoes a further metamorphosis, taking these changes one step further to become what we might call advanced grid infrastructure (AGI), which includes much more than meter data collection and extended remote management from the smart meter. With AGI, real-time system information and control now integrates the functionality of advanced metering not only with DR and outage management and restoration, but also with such DA functionality as capacitor bank control, switch control, volt/VAR control, FDIR, fault detection and isolation and restoration, distribution management, and substation management.

Beyond integration of all these functions, AGI also includes integration at the DER level through inverter management, providing management of solar PV systems, EV charging systems management, and remote ES devices. To manage volt/VAR control of the new devices at the edge, the devices need the ability to make decisions on their own; intelligence at the edge is needed. The utility manages volt/VAR control at the substation level. The 6-kW EV power system will consume roughly the same amount of power in eight hours of overnight charging as the house consumes in a 24-hour cycle. Such a dramatic increase in load will drive the need for edge power management. As described here, AGI functionality ensures that the utility will remain in control of all devices connected to the advanced smart grid, which has the added impact of addressing an emerging business risk for utilities when consumers gain increasing amounts of responsibility for edge device management.

Electricity and Smart Buildings and Appliances: DR

More and more builders are adopting methodologies outlined by the U.S. Green Building Council (USGBC) and pursuing Leadership in Energy and Environmental Design (LEED) status. LEED is a classification scheme developed by the USGBC that provides guidelines and independent certification for builders who focus on unique design to promote not just efficient use of energy, but also more efficient use of water, lower CO_2 emissions, and improved internal comfort for the building occupants. Beyond LEED, electric utilities are promoting more efficient buildings through green building programs that encourage enhanced energy efficiency through new technology such as closed and open cell spray foam insulation and advanced window design, but also a plethora of simpler technological practices such as sealing around doors and windows, radiant barriers in attics, and blankets around water heaters, all of which serve to eliminate energy waste and reduce energy system capacity requirements, collectively postponing system demands to expand capacity by building new generation plants. Another aspect of energy efficiency concerns replacing appliances with newer, more efficient designs that consume less energy, most notably HVAC systems with higher standard energy efficiency rate (SEER) ratings, but also more efficient water heaters, refrigerators, and CFL and LED light bulbs, among other innovations. As with smart buildings, smart appliances reduce energy consumption by substituting more energy-efficient devices for less efficient ones. Finally, moving beyond the built infrastructure of smart buildings and smart appliances, energy conservation is rounded out when the focus shifts to a more active focus on human consumption behavior patterns throughout the day, leading to lower energy consumption throughout the day, but also to more fine-tuned conservation that can lower peak demand for optimal utility operations.

To draw the contrast between DR and energy efficiency further, changes in behavior designed to lower overall consumption fall under the conservation heading. However, changes focused on shifting consumption during specific times of day (peak periods when the costs to produce electricity are the highest) are referred to as DR, when the demand side of the equation—energy consumers—respond to utility requests to lower consumption temporarily to either reduce the production of high-cost energy (i.e., economic DR) or relieve stress on an overburdened distribution grid (i.e., reliability DR).

Looking 10 years into a mythical future, the scenario in the following use case imagines the impact rapidly changing technology will have on the electricity grid, driving the demand for an advanced smart grid and creating an alternative energy economy based on negawatts, or avoided energy production. (Negawatt is a term that we believe was first coined in the 1970s by Amory Lovins of the Rocky Mountain Institute.)

Use Case of the Future: DR in 2025

By 2025, most consumers were well on the path to shifting their electricity consumption away from peak periods, spurred by widely implemented DR programs coupled with new time-of-use (TOU) pricing incentives.

TOU rates required three key changes: (1) an AMI with digital meters producing interval data, (2) a digital billing system that could produce bills that leveraged the interval data, and (3) an analytical study, followed by a TOU rate case.

Once in place, DR programs provided utilities three key benefits: (1) avoided capital expense from construction of new peaking power plants, (2) avoided operating expense from running aging power plants and from purchasing expensive power and transmission on the spot market during critical peak periods, and (3) enhanced profitability through selling relatively more power during lower-cost, off-peak periods.

By early 2020, utilities could begin to analyze the impact of the DR programs they had put in place. In effect, system operators had been seeking a new resource they could dispatch like traditional generation to keep the system in balance. Within a year or two of a DR program's implementation, megawatts of electricity demand would become available for load shedding, growing steadily as program acceptance grew. Increases in rates made DR negawatts more attractive still, leading to hundreds of DR megawatts capacity by 2025, effectively offsetting the construction of a midsized power plant for many utilities.

Another benefit of DR and TOU rates was that more and more demand shifted to match those times when renewable energy came on the system. In early DR programs, notification of curtailment need was manual, through phone calls or Web site postings. However, by the end of 2018, automation hardware and software could send digital signals to user HEMSs and to user communication devices like smart phones and laptops, but also directly to appliances like smart HVAC units, water heaters, and refrigerators to prevent them from consuming power during peak period [such automation had introduced a new term, automated DR (ADR)].

By 2025, the DR system had proven itself sufficiently reliable and predictable to be built into integrated resource planning (IRP) as just another resource to meet system demand. Furthermore, together with efforts to improve energy efficiency, these DR programs enabled utilities to meet their portion of state mandates for peak load and carbon reductions at the lowest cost possible.

Power Engineering Concept Brief

As seen by the glimpse into the future offered in the use case above, DR holds great potential to become an integral part of the system of dispatchable resources under the control of the grid operator. Peak shaving, the earliest goal of DR programs, is a term used to define the reduction of the highest point along the load curve through targeted energy use curtailment. Peak shifting is a more complex form of DR that requires very close correlation with more sophisticated users, where the utility is able to move the peak and the load curve to more closely conform to its most efficient production curve. Carrying this concept still further, when the users are working in complete harmony with the energy producer, automated demand management controllers tailor the load to optimize the physical and economic delivery of energy to eliminate inefficiencies and maximize revenues. The ultimate for electric utility operations is a predictable, flat load curve, so that supply may be effortlessly aligned with demand.

To achieve this vision, HEMSs need to do more than just present detailed feedback on energy consumption to energy consumers. When equipped with automated algorithms, programmed to meet users' unique profiles of comfort, convenience, cost, and carbon (4C), the HEMS devices will be more effective at reducing loads in conformity with utility needs. When the HEMS devices are connected to smart meters through local wireless technologies like ZigBee, they will not only have access to interval data, but also provide an opportunity for utilities to engage directly with the HEMS devices when the customer allows it. Special pricing programs that can correlate with the 4C user profiles using the HEMS system will stimulate user behavior to more closely align with utility needs.

From a power engineering perspective, DR programs will first segment the market along a spectrum of the most willing, flexible energy users down to the least. These programs will present such groups with tailored tools, pricing, and programs that allow them to conveniently adjust their behaviors to meet utility needs. Beyond the DR programs, the tools will include HEMS (and for commercial customers, BEMSs) that direct information to the device of the consumer's choosing. Those devices will need to correlate both with data received directly from inside the premises, with data from the smart meter, and perhaps with data such as pricing information and special deals from the utility delivered through the smart meter (or over the Internet).

Interval data from smart meters, collected and organized with meter data management (MDM) tools, will be mined using data analytics programs to produce information that can feed such tailored DR programs. With the information and insights derived from huge databases, patterns can be detected, causal relationships identified, and buyer groups formed to offer a new variety of energy services that appeal to consumers who grew up with one-size-fits-all commodity electricity.

Consumers and Prosumers: DG

With DG, we see energy consumption converging with energy production. When energy consumers enable their premises to produce electricity using DG technology, be it with rooftop solar PV, microwind, natural gas microturbine CHP, or other technologies, they create on-site power plants that come with a built-in efficiency upgrade: They are not subject to line losses. Such a transition holds revolutionary potential that depends on consumer adoption, which is driven by price as well as awareness. As technology progress makes DG both more productive and less expensive, as traditional electricity prices rise, and as DG becomes less exotic and more and more neighbors opt in, the appeal of going "off-grid" with a DG system will continue to grow.

A rooftop solar PV system is likely to be the predominate form of DG that a utility will encounter for the foreseeable future, given the rapid price declines and the flexibility and maturity of the platform relative to the other technologies. As described in Chapter 2, the principal components of a PV system include: (PV panels and the BOS, which includes everything else, from wiring and switches to inverters, batteries and net meters. For a PV system to become a DG node connected to the advanced smart grid, however, the inverter is the most likely component that will need to be made "smart" by adding a communication capability and localized intelligence. With a smart inverter, the operator in the utility NOC will be able to "see" the asset and then send messages to control it. Such controls would provide direction that could use the energy produced by the smart inverter to achieve such system benefits as volt/VAR regulation, and in peak times, they will provide access to renewable energy that could be marketed into the wholesale power market.

The challenges of connecting thousands—or even tens of thousands—of rooftop solar PV and other systems to the smart grid today remain manageable only as long as the amount of units per distribution feeder stays small, and as long as the energy each DG system produces doesn't flow unmanaged back onto the grid, which would put expensive distribution substation gear at risk—two significant constraints that could severely limit the uptake of DG as grid parity approaches. As with EVs in the next section, the challenge of integrating these DER systems is an issue that should be tackled sooner rather than later. Rather than pondering how many systems might be added to a utility service territory over the coming years before they must act, utility managers should be considering the threat that DG poses to equipment along a single, overburdened distribution feeder and the opportunity that DG offers to engage with customers in new ways. Management capability at the feeder level will be critical to the integration and performance of DG in an advanced smart grid.

The term high-penetration PV (HPPV) is used to describe efforts to load up a single distribution feeder with higher concentrations of PV facilities than current standards allow. A rule of thumb in the utility industry today is that a single distribution feeder can safely handle about a 20% penetration of PV. Go north of 20% and the potential for intolerable risk and instability to grid operations sets a boundary—the intermittency of production and the potential for excess power to reverse the power flow on distribution feeder lines pose a threat to upstream equipment.

Another term, VPP, describes a demand-side alternative to accommodate growth in peak demand to the traditional supply-side alternative of adding a natural gas power plant, commonly referred to as a peaking unit or a peaker. In its most expansive definition, a VPP combines an array of rooftop PV systems with localized ES and aggregated DR capacity (e.g., HEMS appliances equipped with direct load control—commonly, smart thermostats—or some combination). Such a system provides a utility the capacity needed to meet its peak needs without a power plant.

At the micro level, HPPV and VPP require technology to be refined at the scale of a single distribution feeder or neighborhood. Progress is afoot. Two innovative utilities at the forefront of research on HPPV, Sacramento Municipal Utility District (SMUD) and Hawaii Electric Company (HECO), have been conducting R&D projects since 2010 in joint cooperation with the National Renewable Energy Labs (NREL) and coordinating their efforts to provide valuable pioneer research [3] on the challenges and potential solutions regarding HPPV with funds from the California Solar Institute and matching grants. Duke Energy is also testing VPP technology on a small scale at its McAlpine Creek Substation project [4], and the Center for the Commercialization of Electric Technologies (CCET) has a VPP project as a subset of its DOE ARRA Demonstration Grant project [5], where it showcases its Smart Grid Residential Community of the Future with the Pecan Street Project in Austin (see Chapter 5).

Use Case of the Future: DG in 2025

After the conclusion of national elections in late 2020, when Congress passed a landmark climate bill that mandated 85% carbon reductions by 2050, electricity leaders nationwide knew they would need every clean energy resource they could find, which spurred a boom in renewable energy. Small-scale DG, in many ways, was well suited to the challenge of climate control.

DG put residential and commercial energy consumers in charge of their own destiny, with greater self-sufficiency and security, insurance against future energy price shocks, and a chance to take advantage of major new economic opportunities, including thousands of local jobs. The transition to DG started simple enough, but that is not to say that there weren't bumps in the road. Most of the solar PV capacity would end up coming from crystalline silicon and thin film solar PV modules, the leading technology options at the time. Installations were mostly on residential, commercial, industrial, and city-government rooftops, but some small, local solar gardens were also built, many of them pulling double duty as parking lot shades and charging stations for EVs.

The benefits of DG were broadly shared: Besides all the homeowners who installed rooftop systems, residents without rooftops or backyards were able to buy shares in midsized solar cooperatives (500 kW–2 MW). Utilities also rented large commercial building rooftops to deploy utility-owned solar, leased solar modules (and later, ES systems), and bought local clean energy from private developers, packaging DG with low-carbon centralized generation to expand affordable clean energy.

New ES technologies, as they came on-line in the second half of the decade, did more than anything else to spur greater interest in DG, transforming intermittent renewable generation into firm, dispatchable power. When a utility could schedule to buy and store low-cost energy to use at peak, high-cost times, the appeal of DG became ever more clear. The list of utility benefits of stored DG included the ability to: (1) replace spinning reserve, purchased power, or new peaking plants; (2) defer investment in T&D upgrades; and (3) improve power quality, reliability, and outage management programs.

While most discussion about DG concerned solar PV, it should be noted that solar PV did not tell the whole DG story. An economically compelling slice of DG also came from such technologies as natural gas-fueled microturbine CHP units, and the conversion of landfill gas (LFG), waste heat, and waste biomass into electricity. These relatively low-cost opportunities were already at or below parity with fossil fuel generation at the end of the first decade of the new century. While the potential of waste heat and biomass was limited to the availability of waste and biomass, such base load technologies proved uniquely valuable for reducing greenhouse gases. First, every lump of coal avoided because of waste heating and energy production also meant avoided methane emissions, and since methane has more than 20 times the global warming impact of carbon dioxide, waste energy became more and more popular. By 2025, the conversion of waste to energy had become widely valued as an important component for a utility to achieve its environmental goals.

Power Engineering Concept Brief

In power engineering terms, what is a smart inverter, and what is the required functionality to make a solar PV system a dispatchable node on an advanced smart grid? First, let's discuss the current design and functionality of an inverter connected to a standalone solar PV system. The fixed-output inverter is one of the simplest designs possible for an inverter, and the inverter most often chosen for a solar PV system is known as a true sine wave (TSW) inverter, which produces a TSW as the name suggests, providing high-quality power that does not produce adverse effects. These inverters have a principal task: to convert the DC power output of the solar PV system into a steady stream of AC power that can run through a net meter to be measured, then on through the circuit box to be consumed within the premises. If the load inside the building is less than the output of the system, the system sends that power back onto the grid. In effect, the tasks demanded of this inverter are relatively straightforward compared to what we will ask of the smart inverter–equipped solar PV system. For the purposes of this analysis of the smart inverter, let us also assume that the local DG system includes a local ES unit to provide more flexible use of the power produced by the smart solar PV system.

The smart inverter is equipped with two key capacities that separate it from its simpler cousin. First, the smart inverter possesses localized intelligence—business rules embedded on a chip—that lets it make decisions that suit both the needs of the system owner and the needs of the utility. Second, the smart inverter is also equipped with communication capability, either to communicate locally with the nearby smart meter, or to communicate over longer distances within a local or regional network. Ideally, a third key capacity will be added to any smart inverter in the future—power electronics that enable the output of the smart inverter to be tailored to the needs of the system, varying voltage to VAR output as needed.

In one way, the smart inverter acts as a dispatching agent, deciding on a moment-by-moment basis whether the power should be stored locally, fed to the premises to offset grid power consumption, or fed back into the grid. To make such decisions, the smart inverter requires information on the current and historic power consumption at the premises, the charge state of the connected storage device, and the market price and grid's ability at any time to accept voltage or VARs onto the grid.

The advanced smart grid is capable of managing thousands to millions of these new smart inverters, automatically dispatching their power when needed to optimize the grid and to take advantage of market opportunities where renewable energy is priced at a premium. Having power input both from centralized generation at one end of the network and thousands of smaller DG units at the other end provides a radical new capability to grid operators to achieve grid optimization.

Electricity and Transportation: EVs

Transportation and electricity infrastructure have much in common, but historically the transportation industry has been dominated by a dependence on petroleum products, which has resulted in EVs being relegated to tight niche purposes, such as mass transit vehicles, airport vehicles, and forklifts in industrial facilities. Until recently, there has been little chance of the two industries converging when it comes to personal transportation. However, considerable progress over the past decade in the technology, design, and development of a commercially viable EV has changed that, leading to an explosion of interest and the need for all electric utilities to reconsider the potential impact of a massive EV adoption, especially given the impact on specific distribution feeders in potential high-adoption areas of a distribution grid. Convergence of these two foundational industries is now not just likely; it has become a foregone conclusion. Before moving on, a word or two on EVs and charging station technology is in order.

While battery alternatives for EVs range from lead acid to nickel metal hydride, the dominant ES technology has become lithium ion, more specifically, lithium ion phosphate, which seems to offer the most appealing combination of low-weight, high specific energy, and energy stability to make it the most appropriate technology selection for transportation applications. Regardless of the battery technology, however, electricity capacity in a battery is measured in amp hours, with total energy capacity denoted in watt hours. Charging an EV depends on the voltage at the socket and in the charger capacity, including both capacity that is external to the vehicle and capacity in the on-board charger itself. As stated previously, a good rule of thumb is that a typical EV is likely to double the electricity consumption of a typical home. Clearly, the impact of this huge jump in electricity consumption on a localized basis in a distribution grid must be considered by electric utilities well before EVs become widely adopted.

Level I charging occurs at the standard voltage of a typical electrical outlet in the United States: 110 to 120 volts, which can result in a charge period of between eight and 16 hours. Level II charging is more suited to most charging needs, taking from four to six hours at 220 to 240 volts (the voltage typically found in the outlet used for a clothes dryer). When away from home base, EVs are likely to seek a more rapid charge, so Level III charging uses 440 volts, providing an 80% charge in as little as 30 minutes. However, Level III chargers are more expensive, and given their high voltage stress level on the power grid from high sudden load charging, they require substantial training and expensive facilities.

Three key elements provide insight into the convergence of transportation and electric grids: (1) EVs as new electricity demand, (2) EV charging station infrastructure, and (3) EVs as a utility ES resource.

First, EVs represent new electricity demand. Because the fuel source of an EV is electricity, these vehicles represent a significant new load to be added to the electricity grid. The impact of an EV on grid operations depends on the rate of EV charge, the frequency of charging, the time of the charge (e.g., peak or off-peak-), the location of the charging, and the level of charging coordination (such as planned or unplanned). The key question for utilities to answer concerns how to manage this new type of load as a potential burden or threat to operations. If electric utilities can shift this new load to times of day when the utility's generation resources operate with considerable slack, then they can achieve greater capacity factors and efficiencies and improve profitability. However, if EV owners plug in when they arrive home, the collective new EV load during the peak periods in the evenings and during the summer will stress the grid still further and require significant capital investment to make the grid more robust.

Second, as discussed above, EVs require a charging infrastructure, likely comprised of three main charging alternatives: (1) on-premises charging stations for the "home" location, whether a residence or a business (Level I or II); (2) public charging stations, open to transient EVs on a scheduled or ad hoc basis (Level II or III); and (3) private charging stations, closed to public use but open to fleet EVs (Level II or III). The design of the infrastructure will be a critical issue for utility operations, as it will go a long way to determine how and when EVs interface with the grid. The key question to answer concerns the optimal infrastructure design for a utility.

Integrating the emerging EV infrastructure into the utility grid, into local communities, and into individual households will require tremendous cooperation and planning. Utilities will need to roll up their sleeves on this one—they will need numerous trials to refine the details and determine the appropriate vehicle-to-grid (V2G) and vehicle-to-home (V2H) processes and policies.

Finally, EVs represent a potential new ES resource—if one that remains a long way off. The challenge in this regard will be to determine the best way to take advantage of a fleet of distributed storage elements with a significant amount of collective storage capacity. As consumer adoption of EVs progresses, the collective storage capacity of fleet vehicles will grow as well, giving promise for a new clean-tech resource for utility-wide load shifting. Perhaps the most intriguing proposition regarding EVs is that they could be used to store renewable energy production, particularly power from wind farms, which produce most prolifically during off-peak hours (i.e., during the night), when energy demand and energy prices are lowest, and from solar PV arrays, where the production cycle extends only a few hours into the peak period, which could be extended with integrated storage. Using EVs to store wind and solar power for use during off-peak periods could provide significant arbitrage value if that stored energy could be discharged to the grid during peak periods when both demand and prices were far greater. A burgeoning V2G storage capacity will enable smoother utility distribution system operations, and a growing V2H capacity could be expected to provide a significant boost to the potential of DR programs as well.

Use Case of the Future: EVs in 2025

By 2025, the signs of the electrification of transportation systems were increasingly apparent. Roughly 35% of new car purchases were EVs, a category that included not just plug-in hybrids, but all EVs. Incentive programs engaged utility customers and EV dealerships to install the charging station infrastructure to help support the surge in EV ownership. New EV owners, both individuals and fleets, were drawn by the combination of high fuel costs and popular incentives. For Austin, a city of one million, these penetration numbers meant that about 200,000 EVs circulated daily throughout the city in 2025.

Electrification wasn't just about EVs, though. Mass transit had become more electrified too—legacy rail lines used electricity as a growing substitute for diesel-electric rolling stock. Preliminary deployments of EV bus fleets were under way, and in airports across the country, full electrification of ground support equipment had been completed for years.

EV program planners discovered more usage patterns than the nightly charging scenario that took advantage of low rates and coincidence with wind energy production. Some users had to recharge during the day based on their personal schedules; others participated in EV car-share programs, which required frequent recharging between short drives. School buses faced heavy use for nine months, but little to none during the summer, freeing their storage capacity for other purposes. Some workers had chargers at their workplace, and more and more multifamily housing developments and retail establishments continued to add charge stations to make their locations more competitive. The EV charging market developed steadily as third-party companies sited recharging stations throughout the city at strategic locations, most especially high-traffic retail sites, mass transit parking lots, and on the rooftops of parking structures. Finally, city building codes needed to be adjusted to accommodate EV charging, reflecting the more integrated approach adopted by utilities over the course of the decade.

As utilities integrated EV charging capabilities with their smart grids, they gained a new automated ES resource. Charging and discharging were programmed over the smart grid to achieve an optimal balance between grid needs and the needs of the individual EV owner. Dubbed EV support equipment (EVSE), these smart charging stations accommodated the new TOU rates and fed relevant information to digital billing systems when and as needed, in support of TOU rates, real-time pricing (RTP) options, generation fuel-mix forecasts, wind and solar generation signals, and other customer-specific information, providing a range of charging options (e.g., charge at lowest cost, charge for lowest carbon footprint, and charge immediately). The EVSE was designed on industry standards in cooperation with other utilities and the EV industry to ensure interoperability, flexibility regarding communication types, and low-cost production. By 2025, the relationship between the consumer and the utility had evolved considerably, and the complexity was managed by ever more capable technology and automation.

Power Engineering Concept Brief

EV as Stationary Load

In the utility world, load or system demand needs to be managed and planned, and that makes the unpredictable nature of EV adoption rates a huge challenge.

For the sake of clarity, let us focus first on the management of stationary charging stations and the load they will place on the utility, whether they are located in a residential customer's garage, on the curb, or in the parking lot of a small business. If for no other reason than to ensure that the utility will not take an unplanned risk with regard to mobile EVs, it is essential that system planning to manage charging station loads be coordinated with these different customer classes. Utilities will need to develop a strategy that targets specific distribution feeders for upgrades to harden them against the risks of overloading from too many EVs charging at once.

EVs as Roaming Load

Mobile load represents a unique management challenge to a utility—never before encountered given that providing power to meet demand has heretofore been relatively predictable with regard to any grouping of grid termination points. Thus, utilities may also consider the relative costs and benefits of two strategies seemingly at two ends of a spectrum of options: either corralling and clustering EVs when they are roaming by concentrating charging stations (e.g., providing charging stations to garages and parking lots), or distributing and dispersing charging stations throughout the territory to defray the impact they may have on the system at any one time. Of course, planners may choose to do one strategy before the other or to do both simultaneously. The idea is to design a charging system infrastructure that drives EV charging activity to an outcome that meets societal needs and conforms best to the needs of the utility.

Examining how utilities have managed the deployment of rooftop solar PV systems is instructive to discern how they might approach EVs and related charging infrastructure. So far, most utilities have let the market control where solar PV systems are deployed—perhaps offering rebates to defray costs, perhaps under rules and guidelines for system connectivity (i.e., net meters), but generally taking a hands-off approach. Some utilities may choose a similar path for EVs, perhaps predicting that the pace of change will be slow enough to manage the situation and learn from the outcomes. The benefits of such an organic, market-based approach are to position chargers where they have most value early on and to help define future infrastructure investments that will match the market requirements. In this approach, costs lag EV adoption; this could be termed a deferred investment strategy. If the utility is not sufficiently proactive and underestimates penetration, however, it will find itself in a reactive mode, challenged to manage the EV-related peak load, and risk instability.

Conversely, a utility that takes a more hands-on approach with PV by proactively locating community solar PV systems, investing in rooftop leasing programs, educating the public on the dos and don'ts of solar PV, and enabling new market opportunities to emerge has, in essence, partnered with the market to guide outcomes more favorable to its long-term interests. Utilities may do the same thing with EVs and face higher initial investment; costs will include new EV charging stations across the planned locations, new power electronics to accommodate new load concentration, and new customer support programs. The benefits of such proactive planning for the location and adoption rate of EVs include increased coordination with daily operations and lower long-term risk.

This proactive scenario is harder than it looks, however. First-generation EVs adhere to certain standards, but they are not yet integrated as dispatchable assets under the control of the utility; the utility is not able to control with any specificity when and where the EVs charge. Utility investment in charging infrastructure must perforce be a strategy of control and management of outcomes. In this approach, the utility subsidizes the charging equipment, not unlike the cable companies subsidize the DVR, to stimulate market adoption, but also to generate ancillary revenue and for operational control. Managing a charging infrastructure will require the utility to address such issues as location, design, and functionality of the charging stations; customer identification, metering, and billing; and the new power electronics needed to enable such functionality and control.

To grasp the significance of EVs to utility operations, imagine a group of friends driving to Austin, Texas, to attend the 2020 South by Southwest [6] (SXSW) Interactive, Film, and Music Festivals during spring break in mid March. Heading to Austin from Seattle in her "extended range" EV the driver enters "SXSW" into the navigational system. The EV maps the best possible route to a series of chargers along the route, based on the driver's choices (e.g., cheapest charging, cheapest hotels, best restaurants, and best scenery). As the friends start their journey, the EV has already contacted Austin Energy to open an account (or reactivate an old one), providing the driver's information (such as name, address, telephone, and credit information) and prenegotiating the best rates and locations to charge the EV while in town, based on the driver's preset condition parameters.

While the driver and her friends are parked at SXSW enjoying the festivities, the car becomes a provider of energy and energy services to the grid, generating revenue for the driver (one hopes sufficient to offset parking charges!). The car serves as an on-demand capacitor to the local grid via the "SXSW-EV Program," which pays a premium above regular rates during shoulder peak and full peak hours to ensure that the utility can manage the situation and to gain benefit from the influx of EVs, which are in effect mobile storage units—DER devices. Such coordinated programs will optimize benefits for drivers, utilities, host chargers, and local retailers.

Electricity and Warehousing: ES

At the risk of stretching this convergence meme as far as it will go, we come to the end of this chapter with a discussion on the introduction of ES into the electricity supply chain, which must be viewed as the "mother of all game changers" for the utility industry. Consider that throughout the power supply chain, the rules of the game have forever been written around the requirement that the system operate in real time because of the lack of economic ES alternatives. There has never been much of a warehouse alternative in this particular supply chain.

We would like to propose that hydropower was the very first and closest thing to a feasible, widely adopted kind of "warehouse" alternative, the electricity supply chain developed as if ES would always be: (1) very expensive; (2) difficult to site, ruling it out in most instances; and (3) rather clumsy in its application, not providing the fine-tuning one would hope for in a storage asset. Therefore, it should come as no surprise that the electric system developed as it did over the next 100 years, designed and operated according to assumptions such as the following:

1. The system must be kept in balance;
2. Most load is unchangeable, so we must get good at following load with generation to keep the system in balance;
3. The primary challenge is to manage the system capacity to ensure availability during peak consumption periods;
4. Supply-side resources are the dominant solution to add significant capacity to the system.

Where grid managers and designers could employ hydropower as an energy resource, they did. The first significant power plant was a hydroelectric dam between Niagara Falls and Buffalo, New York. Indeed, the early history of electricity is very much concerned with acquiring rights to rivers and building hydroelectric dams. Hydropower, in fact, has subsequently been adapted into a particular ES technology referred to as "pumped hydro," where smaller paired reservoirs are constructed for the specific purpose of ES—using electricity to pump water up when electricity is cheap, then letting gravity drop the water through the system to generate electricity when market prices are elevated.

Use Case of the Future: ES in 2025

For those utilities that added utility-scale storage to their distribution grids, the utility operational model that had worked for over 100 years had been turned upside down. By 2025, several storage technologies had reached well into commercialization stage, and price points were coming down, though prices still remained too high for many utilities. Even in 2025, many utilities remained paralyzed by the diversity of the storage technology options available and the rapid changes.

Those utilities that did invest in ES had more options, because now energy could be economically stored and used when it had greater value and more utility. Finally, like nearly every other industry, the electric industry had a warehouse capability. How did they use it? Clearly, utilities were still in the experimental phase. The 2009 ARRA legislation, more commonly referred to as the Stimulus Bill, had primed the pump back in 2009, when billions flowed through the DOE into utility programs, notably the storage demonstration projects under DOE FOA 36 [7].

Leading companies emerged for the different types of ES, and clear leaders for each type of application emerged as well. Managers at utilities had their preferences. Some technologies were simply better for some applications. Utilities had moved slowly into ES, but once a utility found one or more applications and technologies that were market-ready and fit their needs, they made rapid progress. In the early days, utilities used pilots to investigate the possibilities, from combining smaller ES systems—"community ES"—with solar PV panels in a neighborhood to placing larger, utility-scale ES facilities on industrial sites for load-shifting to avoid peak consumption, to using ES to relieve congestion at strategic points in the grid, to collocating ES with renewable energy farms to provide a buffer against disruptive intermittent power production.

Power Engineering Concept Brief

In this section, we are more concerned with ES as it converges with the role of warehousing in a supply chain than we are with individual ES technologies. Our focus is thus on the supply-chain impacts of ES and on the smart inverter, the point of connection between the storage technology and the advanced smart grid. Considering ES from the power engineering perspective, the price of ES in the near term will likely limit its deployment to locations where such an asset provides the most value, whether to accomplish a business goal or to provide information and insight in a pilot or research project.

ES has great potential as an element to transform the design of the advanced smart grid. ES devices will need to be integrated into the advanced smart grid with the following issues in mind. At first, integration will be accomplished in phases, with ES systems added incrementally while ES technologies become more and more economically feasible with dropping prices. Thus, locations on the grid will be targeted based on some combination of economic and system engineering benefits. Second, ES will be considered as a critical element in disaster recovery, which indicates collocation with shelters and critical facilities. Finally, ES will accelerate the processes and changes described in this chapter—when ES is added to the grid or to a DER element, the different components, applications, and design elements of the advanced smart grid described in this chapter become more efficient and versatile, from DG to DR, from EV charging stations to DA. Put quite simply, ES is the most versatile technology when it comes to grid operations and enhancements, so over time it will dramatically reshape the potential of the grid.

Imagine for a moment the load management strategies that grid operators could achieve if they were to deploy ES devices across their service territory to accelerate outage restoration times, minimize load congestion zones, optimize large disaster restoration zones (e.g., schools and churches), improve small disaster recovery zones, and improve all around volt/VAR control, all while improving quality performance indices such as SAIDI, SAIFI, and CAIDI.

In particular, let us examine in more detail one particular kind of ES that is sometimes overlooked. Thermal ES represents hidden potential: For instance, refrigerators and freezers, ubiquitous in households and many businesses, store cold air to keep food fresh. Looked at in a different way, though, these devices are also microwarehouses of thermal energy, which when aggregated, may serve as a resource on the advanced smart grid. During peak periods, such connected devices may be signaled to switch to a conservation mode that will postpone their regular chilling cycle, so that they become a distributed resource not currently in play on the grid. These distributed thermal ES devices change the way we look at storage and appliances. Integrating such assets need not be about any loss of comfort or convenience either, rather tapping such a resource merely involves a minor sharing of a thermal storage resource in a collective strategy to incorporate a new resource that did not exist before. The advanced smart grid will make this possible by providing a network, a network management system, and the smart devices—the smart inverters—as the missing elements to access distributed elements and realize their hidden value.

The key to realizing these and other scenarios lies in the smart inverter, which when connected to the distributed ES device transforms it into a smart grid element. The smart inverter, like the smart meter and the smart router, has a processor to provide local intelligence and communications capabilities (Ethernet/LAN/WAN connectivity), and remote control potential. These three attributes enable management of massive numbers of end devices in an advanced smart grid.

Conclusion

In this chapter, we highlighted two megatrends that are transforming infrastructures as diverse as the electric grid and the state highway system: digitization and networking. In fact, all infrastructures have the opportunity today to add networked digital sensor and control devices to gather information on infrastructure status and operations, whether the commodity they move is electrons or vehicles. In short, all infrastructures benefit not only from access to such revolutionary digital technologies, with new devices emerging every day, but also from advances in network technologies that enable that information to flow back to management consoles, databases, and servers, where the data may be acted upon with new data analytic software to provide insights never before available. These trends make formerly diverse infrastructures more alike and lead them to work more closely together, even interoperating in some cases, as when water systems automatically curtail their usage during peak electricity periods to save the energy that would be used to pump and treat water during those critical times. We've labeled these activities using the term smart convergence, where the infrastructure managers learn from each other and where possible, leverage each other's infrastructures to achieve still greater operational efficiencies.

Infrastructures that benefit from smart convergence share the following core functions: (1) distributed, which have elements they draw upon throughout the infrastructure; (2) interactive, where the elements of the infrastructure interoperate and influence each other; (3) self-healing, where the elements work together in such a way as to promote improved performance; and, finally, (4) ubiquitous, in which their converged qualities are found in every device.

Smart convergence, as it is recognized and employed by infrastructure operators, will have significant consequences for all aspects of our modern economy, which rises and falls based on the success and health of its multiple infrastructures.

Smart convergence will lead to dramatic cost reductions when infrastructures share common elements, but also to dramatic increases in effectiveness when a combination of infrastructures leverage efficiencies or when borrowing best practices, improving operations and changing potential by recognizing and incorporating new assets heretofore unavailable.

In Chapter 4, we will trace the origins of the advanced smart grid concept in an extended case study of Austin Energy, starting in 2003 with reforms taken to make the IT back office more efficient and cut costs, leading through a variety of projects to address application silos and integrate new applications and smart devices, and finally resulting in the emergence of a pioneer utility-wide smart grid in 2009. The seven-year process proved both valuable and instructive, achieving its stated goals, but also revealing lessons learned and insights on smart grid through both its successes and failures.

Endnotes

[1]   http://www.utc.org/.

[2]   http://www.thocp.net/hardware/atm.htm.

[3]   http://www.solarnovus.com/smart-grid-leveraging-the-disruptive-impact-of-solar-pv_N1867.html.

[4]   http://www.duke-energy.com/news/releases/2009061602.asp.

[5]   http://www.electrictechnologycenter.com/doe.html.

[6]   http://sxsw.com/.

[7]   http://energy.gov/sites/prod/files/ESS%202012%20Peer%20Review%20%20DOEOE%20FY12%20Electrical%20Energy%20Storage%20Demonstration%20Projects%20-%20Dan%20Borneo%20SNL.pdf-

CHAPTER 4

SG1 Emerges

Chapter 3 explored the different infrastructures that we depend upon for our modern economies and lifestyles and showed how they are converging on themselves. The electricity grid in particular is incorporating telecom practices and habits, and IT is becoming an ever more vital aspect of electricity grid operations. In Chapter 4 we use a case study to showcase and evaluate such concepts in a single utility, examining in detail an actual SG1 implementation. The case study approach reveals any number of lessons learned, but also details the emergence of the advanced smart grid vision, showing how it developed through trial and error in a real-world living laboratory environment.

Introduction

This chapter describes the genesis and implementation of a smart grid at Austin Energy, the eighth-largest city-owned electric utility. Starting in 2003 when the term "smart grid" had barely been circulated among utility cognoscenti and had certainly not gained the widespread acceptance it enjoys today, we defined what is today called a smart grid through the series of incremental steps we took at Austin Energy described in this chapter. The smart grid as it came to be defined at Austin Energy was more expansive than other definitions at the time (e.g., EPRI's Intelligrid, IBM's Intelligent Utility Network, and Meta's Geodesic Energy Network), which were centered on the utility electric infrastructure alone. Starting in 2004, when Austin Energy's CIO Andres Carvallo first started talking publicly about a smart grid, he combined the electricity infrastructure owned by the utility on one side of the meter, with infrastructure beyond the meter owned by customers, from private and publicly owned buildings to individually owned EVs, to DER and smart appliances.

This chapter describes the vital first step of *technology infrastructure rationalization*, critical to the future success of any smart grid project. It explores the creation and adoption of a *technology governance framework* and *smart grid architecture*, as well as the expansion of *networking assets* (fiber to every substation, wireless AMI system deployment territory-wide, wireless DR system extended to smart thermostats, wireless DA system reaching a limited number of sensors for grid optimization), and the addition of a variety of applications through over 150 separate *systems integration projects* to create the first smart grid. Finally, the chapter describes how all this work led to the insights for a new approach to building a smart grid, which we have labeled an advanced smart grid as described in Chapters 1–3.

As will be demonstrated by the case study below, there were two key lessons learned that deserve highlighting. The first concerns the importance of preparing the IT department as a firm foundation for the smart grid to follow. Without a rationalized technology infrastructure, the complexity of a smart grid project guarantees substantially higher risks and costs, even so high as to put the project itself at risk. The second concerns the revelation that beginning a smart grid project by acquiring network capabilities, instead of by incremental addition of applications, provides immeasurable benefits, as described in Chapters 1–3.

The authors of this book, Andres Carvallo and John Cooper, collaborated twice in projects at Austin Energy as it was building the nation's first smart grid deployed over a full utility environment. Andres Carvallo served as CIO of Austin Energy from 2003 to 2010 and was the visionary, principal executive champion, and chief architect of Austin Energy's smart grid. Andres Carvallo hired John Cooper for six months to work as a utility applications and IP network communications consultant for Austin Energy in 2004, and then later in 2009 for eight months as a project manager of the smart grid team within the Pecan Street Project, described in full detail in Chapter 5.

Case Study: Austin Energy, Pioneer First-Generation Smart Grid

This chapter tells the story of how the very first comprehensive, utility-wide first-generation smart grid came to be built in Austin, Texas, at Austin Energy, the city-owned electric utility that serves over 1.2 million residential end users and 43,000 industrial and commercial businesses, distributing electricity to over 410,000 meters. The lessons learned in the smart grid journey described in this case study are fundamental to understanding the concepts in this book, which derive not only from the successes and lessons learned in Austin, but also from our work in Texas and in the United States from 2003 to 2013.

To enhance clarity and provide greater insight, we present the remainder of this chapter in a case study format, as told from the perspective of the CIO during that time, Andres Carvallo.

Saying Yes to Opportunity

Juan Garza, the general manager at Austin Energy, sought my help in January 2003, to leverage technology to transform the city-owned utility into a new kind of organization capable of greater flexibility, so it could adapt to the changes that were sure to come. A future journey of personal and organizational transformation began with a phone call requesting an interview (Juan had heard I was recently back on the job market). Juan told me about his ongoing search for a leader to manage IT inside the utility, describing a difficult situation: Austin Energy was not efficient in the use of high-tech resources, Juan told me, as he detailed the lack of tools, reports, and visibility that frustrated executive decision-making and management of the enterprise. Juan and his executive team needed better processes and systems and better integration between IT and other parts of the utility.

That morning, Juan described to me a utility that is not unlike many in the developed world today. The information technology and telecommunications (ITT) department was responsive to the other departments at the utility, but its best efforts were too often stymied by events and constraints seemingly beyond their control. The budget was forever inadequate—demand exceeded supply. More than 90% of any annual IT budget was used to maintain current systems (also known as "keep the lights on"), leaving little to no money to finance the strategic projects that promised to lift the department out of its daily frustrations and deliver the capabilities sought by the other business units that were clients of the ITT department. Juan described an ITT organization that was not working well, providing little information for executives to manage the enterprise.

Before going further, Juan asked me to meet with some of his staff to get some perspective. I met with several of his direct reports, including Austin Energy's CFO Elaine Hart and VP of government relations Roger Duncan, who oversaw all government relations, renewable energy, and energy efficiency programs at the time. I asked them both a series of questions about their use of technology, reports, and systems and how they ran their businesses, among others. These one-on-one sessions helped me get a better idea of what was happening, and I returned to share my assessment with Juan. Citing an array of problems, I confirmed the challenges Juan had described to me when we met—aging infrastructure, inadequate systems, lack of project management, lack of enterprise architecture, lack of even a single version of the truth, and no KPIs used for benchmarking performance.

Juan asked me for a price tag and a time frame estimate to fix the enterprise, and I gave him a ballpark estimate of spending an additional $50 million, over and above existing technology operating and capital expense budgets, to save $100 million. The central concept we discussed was to invest more in technology to achieve savings in capital and operations expenses that would provide sustainable efficiencies across the enterprise. The annual IT budget increased significantly during my tenure, but the corporate numbers tell a more meaningful story. When I joined Austin Energy in 2003, we had about 1,500 employees; 320,000 meters under management; annual revenues of about $750 million; and no online services of any kind. By the time of my departure in early 2010, staffing had grown to about 1,700 employees—about a 12% increase from 2003—but customers and annual revenues had grown comparatively more, to 410,000 meters and about $1.2 billion in annual revenue—about 30%—and smart grid plus full online services had been built.

This then is the story of the smart grid in a nutshell: how we used technology to improve services and enable lean growth. This case study captures the success of a pioneer smart grid journey I was fortunate to have led.

A Fresh Start

In 2003, Juan asked, "Well, do you want the job?" "What is it we're talking about?" I countered. "You come run technology, and we'll invest to free up OpEx and CapEx. We're moving toward these new concepts—DG, energy efficiency, even EVs, and we need to get our house in order."

I accepted the offer, and on February 18, 2003, I walked into my new office for the first time, when I was introduced to a strategic team of executives and senior managers. From the outset, we worked together as a team to redefine the values of the company, including a new mission statement: *"To deliver, clean, reliable, affordable energy and excellent customer service."* I was fortunate, as I said, to come in at a time where there was significant transformation already under way and technology was accepted as a means to an end, helping to integrate new concepts, systems, and processes.

In their utility vision, Juan and Roger saw a bright line between an old way of doing business and a new way. A couple of years into my new job, Juan's vision was finally realized with the completion of a reorganization around two divisions, with Bob Kahn as deputy GM of administrative services and Roger Duncan as deputy GM of distributed energy services. A few years later, Bob left to become the CEO of ERCOT, the independent system operator for Texas, and Mike McCluskey stepped up to become the deputy GM of centralized energy services and chief operating officer. In 2008, Roger would succeed Juan as GM, when Juan left to lead our neighboring utility, Pedernales Electric Cooperative. During the transition, most of the resources remained on the centralized side of the house, but the distributed side grew rapidly, playing catch-up as it integrated new systems and technologies.

As for the ITT department, we had the task of working with both halves of the utility to ensure reliability and continuity of service for current business operations and seamless integration of new functions to the degree possible. In important ways, the ITT department helped to lead the utility in its transformation while I was there, but in other ways, as I said earlier, the transformation to a "utility of the future" vision was already well under way, and our department had to keep up with my innovative colleagues to enable the landmark nation-leading programs they were rolling out, such as GreenChoice, the green energy power purchase program, the Green Building program, and the PowerSaver Free Thermostat program.

Initial Assessment and Issue Identification

I set to work in the first 30 days with an initial assessment. I met with my staff and began interviews to understand technology service delivery and challenges. I also met with a variety of line managers on the operations side—ITT's internal customers—the better to understand their roles and how they used technology and IT services. Those interviews began with owners of customer care, marketing, and finance, and soon thereafter, with owners of operations technology (OT) including electric service delivery, power generation, and wholesale trading.

At this point, it's worth a pause to compare OT and IT, because this relationship is critical to the success of any smart grid transformation. We're all very familiar with IT, but the term OT has less circulation. In the utility, OT describes the power engineering and operational technology groups that manage the generation, transmission, and distribution of power. OT generally is used to refer to wholesale trading operations, as well as power plant engineers, the T&D department, line crews, and metering teams. In a smart grid transformation, it is critical that the IT staff engage seamlessly with the OT staff to upgrade and transform the energy platform to include new information and communication technologies and process innovation. In fact, I would now venture to suggest that OT and IT would be better located together under one executive to truly achieve success.

After those initial meetings, I decided a deeper assessment was in order, so I picked two business analysts in my team to help document the assessment. We set up about 500 one-on-one interviews (out of close to 1,500 total employees) and interviewed each one using a common questionnaire, asking questions such as the following:

- "What do you like/dislike about technology?
- What works/doesn't work well?
- What systems do you use?
- How well do they work?
- Is there a replacement strategy for those systems?
- How do you get support?
- What happens when things break?
- Are the systems internal or customer-facing?"

The results of that questionnaire provided data that let us produce a systems inventory of all technologies currently in use, from the perspective of the users. As we analyzed the interview responses, it soon became clear what was working and what wasn't. As Juan had suspected, our analysis identified a number of fundamental problems.

Issues identified in the initial assessment fell in three major categories. First, the requirements on the ITT department were out of balance with the available resources, principally due to the large number of legacy programs and a lack of coordination and synergy. Second, the necessary IT tools and processes to run a first-class organization were lacking. Finally, the risks that such complexity and disorganization represented to the utility were inappropriate for the mission critical nature of the utility.

Legacy technology systems (Figure 4.1) had accumulated at Austin Energy over the prior 15 years, which posed a dual threat to the enterprise. First, the cumbersome, complex systems caused great frustration for business units that lacked access to the appropriate data to make decisions but were still expected to support increasing maintenance costs. Second, for a provider of mission-critical services like a utility, this technology situation created a significant vulnerability in the form of "single points of failure." The wide variety of systems meant that often a single person would have the responsibility and requisite expertise to ensure the operations of a particular application or system. When that person was unavailable, whether it was single sick day or a planned two-week vacation, the system would also risk becoming unavailable if something were to happen. The analysis documented over 31 mission-critical single points of failure, with potentially fatal disruptions to the utility's daily operations.

Beyond the complex legacy systems and the problems they posed, the analysis highlighted inadequacies in the technology infrastructure that would prevent it from supporting the long-term vision and objectives of the utility. Telecommunications assets in the field and in the corporate offices would need to be upgraded to accommodate advances in digital applications and platforms. Likewise, technology assets in the back office were not set up to manage the massive amounts of

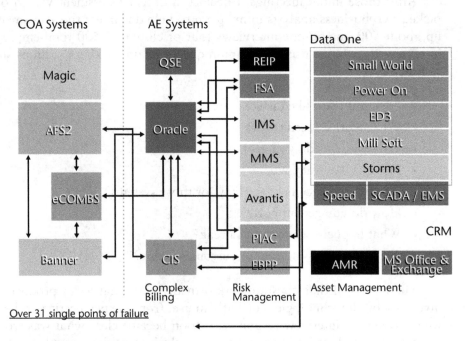

Figure 4.1 Austin Energy legacy technologies in 2003.

data that would be coming their way in the next few years from the addition of a plethora of new data gathering sensors and appliances. Anticipating and preparing for the additional requirements that would be put on the entire organization as a result of massive increases in data became a key driver in preparing for our smart grid journey.

Technology Recommendations, 2003

Based on this thorough analysis, we examined the recurrent themes and with buy-in from other utility executives and created a set of recommendations that became our road map for 2003 and beyond. The recommendations fell in the following four broad categories:

1. Coordinate
   a. Coordinate the purchasing of technology company-wide.
   b. Coordinate all IT resources to improve service levels company-wide.
2. Simplify
   a. Reduce the number of languages and applications supported company-wide.
   b. Automate key missing processes and integrate with legacy.
   c. Deploy a portal for business managers.
3. Expand/upgrade
   a. Expand network architecture to support e-commerce and any device access.

    b. Upgrade data centers, security, and disaster recovery plan.

    c. Implement company-wide smart grid architecture.

    d. Build enterprise data bus and data warehouse/data marts.

4. Invest

    a. Invest in quality, documentation, and training.

By early 2004, the picture at Austin Energy was becoming much clearer. In fact, I had gathered sufficient information by that time to produce a technology strategic plan. The value of a strategic plan goes beyond mapping out a vision, to mapping out the frameworks and goals, and extends to communication of that vision within the organization. The ITT department used the recommendations that came out of the initial assessment and the utility's strategic plan to create five key initiatives for the technology strategic plan: (1) Create and empower a technology governance structure; (2) upgrade and standardize enterprise technology architecture; (3) implement project and resource portfolio management; (4) improve technology alignment company-wide; and (5) increase operational efficiency and quality company-wide.

Accidental Versus Deliberate Smart Grid Architecture Design

Let me diverge for a moment and talk about an important topic to me, one that proved critical to our success at Austin Energy: namely, smart grid architecture, the root of successful transformations. Smart grid architecture includes four levels: processes, applications, data, and infrastructure (networking, computers, and data storage). It is the critical component to make an enterprise flexible enough to adapt to a dynamic and changing marketplace. At Austin Energy, we were fortunate to have an advance view of the oncoming future, given that we were progressive leaders in a variety of areas (such as energy trading, green building, energy efficiency, and green power). We knew that the electricity space was changing into a more dynamic environment, which led us to focus on smart grid architecture as a basis for our transformation.

In my humble opinion, process innovation is the ultimate competitive advantage and the key to any successful transformation. Consider for a moment two hypothetical utilities, identical from the outset in territory: resources and talent. The only way our two utilities in this hypothetical scenario can differentiate themselves in the marketplace is by the way in which they go about delivering their products and services; in other words, it is processes that are the key differentiator. However, beyond the processes themselves, arguably it is their respective *approach* to managing process innovation that actually separates them and determines their different outcomes. In short, the difference lies in *how the utilities innovate for process change.*

And process innovation is supported by, depends upon, and is enabled by smart grid architecture. Rigid smart grid architecture begets slow, limited process innovation. Conversely, flexible smart grid architecture fosters rapid, nearly unlimited process innovation capabilities.

A core aspect of process innovation is focus on those transactions that collapse cycle times, improve customer experience, and enhance customer loyalty, so, it follows that the customer must be an integral part of the transformation effort. Traditionally, that has not been the case among utilities, whose systems have been more concerned with their infrastructure than with the experience of their customers. This focus on infrastructure over customers, by the way, is not limited to utilities; many companies in many other industries focus primarily on applications, data, and/or infrastructure architecture to gain incremental improvements. As we've said, however, it is investment and attention to process innovation, not just applications, data, or infrastructure investment, that leads to profitable and long-lasting improvements.

The electric grid challenge is to move beyond a traditional focus on internal applications, data and infrastructure—the electric grid, as it were—and to expand the focus to include the customers and their systems, the points of interaction within the systems, and the processes that enable transactional improvements.

Let me give you an example of how this worked at Austin Energy. At Austin Energy, we had a single-domain, network architecture—a very strong shield that allowed nothing in or out of Austin Energy—no Internet access, no remote e-mail, no e-payments, and so forth. Essentially, for security purposes, we had an intranet sealed off from the rest of the world. We had network architecture that some would describe as "hard and crunchy on the outside, and chewy and soft on the inside." However, what may be delicious in a brownie is not so good in network architecture. Four use cases demanded that we adjust our network and security architecture.

The first use case concerned allowing utility employees to access the Internet from within the enterprise to do better research and keep up with the new information services online. The second use case was about the requirement to enable remote e-mail and work file access for the employees when they were outside the office while at home or traveling anywhere. The third use case addressed the need for customers to access their usage information and to pay bills online from home or any place in the world. The fourth and final use case, and the most complex scenario, was to allow for the secure access to customer files by customers who are also employees, while they are at work. The challenge of this last use case is that employees start from a secure environment and use company equipment to go out to the Internet, only to come back in to look at their own energy use information and pay their bills, just as a nonemployee could. To do that, however, the network architecture needed to be designed with that particular use in mind. To maintain security and enable all these use cases, we needed to evolve the network architecture to become a multilayer, multiaccess, profile-driven architecture.

We started our evolution by focusing on customer-driven use cases. We had to begin with the customer experience first, which led us to the particular processes that enable those experiences. This wasn't easy, because the cultural attitude inside the utility had been "security trumps everything." We surely recognized the need to maintain security, but we also had to accommodate the demand of new requirements from customers and employees for data access and new services, as we described in the use cases above. Our use case approach led us to processes and requirements and new choices on how to design the applications, data, and infrastructure that would support the processes.

The architecture methodology and design approach that we chose came from the Open Group, a nonprofit organization that specializes in the design of enterprise architectures [1]. Joining the Open Group in 2004 was a critical first step that led to many benefits down the road.

Unfortunately, choosing to design smart grid architecture was not yet a common practice at the time. In fact, we were breaking new ground as pioneers. We deliberately designed our smart grid architecture based on our use cases and the processes we would need to support it. In contrast, the typical path to an enterprise design has been totally accidental, even unconscious, as described in Figure 4.2. Starting with the purchase of an enterprise application, the architecture choice gets made by default. Then the second application purchase repeats this process, potentially resulting in a second and totally different default architecture choice. Repeat this process enough times, as is inevitable over months and years, and the foregone conclusion is a complex management challenge like the one I faced when I started at Austin Energy.

The real need for smart grid architecture becomes self-evident when you consider the operational choices a utility makes as well, revealing a link between IT and OT. This diversion is noteworthy because it describes the genesis of the smart grid architecture that we ultimately created. Somewhere in the middle of these events, I had an epiphany. There was an aha moment for me, when I recognized patterns from other industries and other experiences I'd had in the past at companies like Philips Electronics, Digital Equipment, Borland, and Microsoft. My experiences at companies in the telecommunications and computer industries helped me to reach the enhanced understanding of the true importance of smart grid architecture and the connections between IT and OT that I've described in this section.

It was our adoption of the use case approach that led us to recognize that we needed to move beyond IT and take our lessons learned over to the OT side of the house. We needed to look at operations and segment domains to create a layering

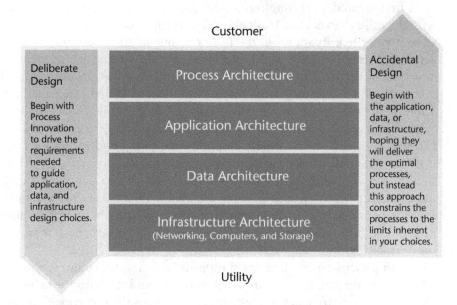

Figure 4.2 Deliberate versus accidental smart grid architecture.

of the security architecture. For example, we needed to isolate critical engineering systems such as SCADA/EMS, OMS, and DMS and provide special privileges and access. Also, we needed to isolate the billing system, the asset management system, and the financial systems. We went from having a castle with a moat and wall, a relatively simple and secure system—but one that was very limiting—to a multi-layer, multilevel, profile-driven access architecture that provided the flexibility we needed to create a smart grid, as shown in Figure 4.3.

Data Flow

A natural follow-up task after crafting our new smart grid architecture was to map the necessary data flows to integrate mission critical utility systems. Such integration is enhanced and optimized by the new smart grid architecture to become a treasure trove of benefits that help make the utility operate more efficiently, increase accountability throughout the utility, and reduce operational and capital expenses by eliminating duplication of effort, manual entry, and paperwork. As stated earlier, the benefits trail starts with a thorough understanding of where the data comes from, how it moves throughout the organization, and the implications of such data flows for technology infrastructure planning and management (Figure 4.4).

Executive Buy-In and Technology Governance

Beyond the initial assessment to highlight critical problem areas, the early technology realignment process depended on gaining buy-in from all departments. Without universal buy-in and a common perspective throughout the enterprise, we weren't going to go far. To execute on the list of recommendations above, I sought to get everyone on the same page by communicating the technology vision and leveraging inclusive processes. To begin to instill order out of chaos, however, it was critical that we create a technology governance plan.

I first created a "technology leadership team" to ensure good communication with the utility's executive team and provide the necessary oversight mechanism we so sorely needed. In addition, the business unit steering committees we established acted as "best-of-breed" decision-making panels, so that departmental executives and leaders had greater control over their functional areas, as well as closer communication and control of the technology resources dedicated to them. The creation of an "enterprise architecture council," a "technology security council," and a "disaster recovery council," with company-wide representation and managed by IT staff, provided the necessary research, planning, and technology selection options to easily meet business unit requirements. The councils were critical to begin the journey of evolving from technological anarchy to practical standardization, increased productivity, and proactive control. Finally, the early establishment of a PMO provided still further structure to allow systematic improvement toward project execution and rational project portfolio management that was consistently on-time and within budgetary limits.

With a focus on creating the standards, policies, procedures, and guidelines by which architecture, security, disaster recovery, and data decisions would be made to achieve great efficiency and effectiveness across the enterprise, the new technology governance plan started delivering at faster and faster speeds the corporate

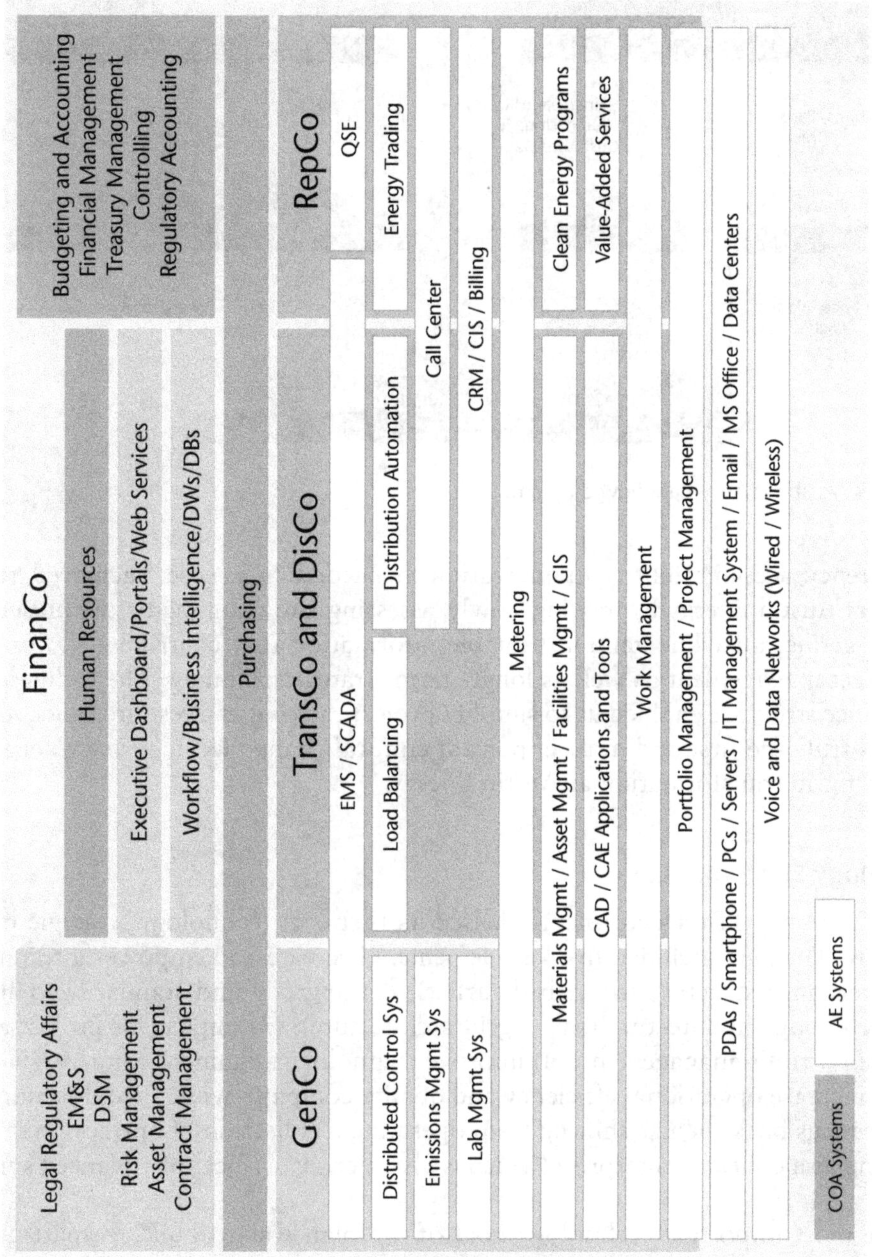

FinanCo
- Legal Regulatory Affairs
- EM&S
- DSM
- Risk Management
- Asset Management
- Contract Management
- Human Resources
- Executive Dashboard/Portals/Web Services
- Workflow/Business Intelligence/DWs/DBs
- Purchasing
- Budgeting and Accounting
- Financial Management
- Treasury Management
- Controlling
- Regulatory Accounting

GenCo
- Distributed Control Sys
- Emissions Mgmt Sys
- Lab Mgmt Sys

TransCo and DisCo
- EMS / SCADA
- Load Balancing
- Distribution Automation
- Call Center
- CRM / CIS / Billing
- Metering

RepCo
- QSE
- Energy Trading
- Clean Energy Programs
- Value-Added Services

- Materials Mgmt / Asset Mgmt / Facilities Mgmt / GIS
- CAD / CAE Applications and Tools
- Work Management
- Portfolio Management / Project Management
- PDAs / Smartphone / PCs / Servers / IT Management System / Email / MS Office / Data Centers
- Voice and Data Networks (Wired / Wireless)

- COA Systems
- AE Systems

Figure 4.3 Austin Energy smart grid architecture in 2010.

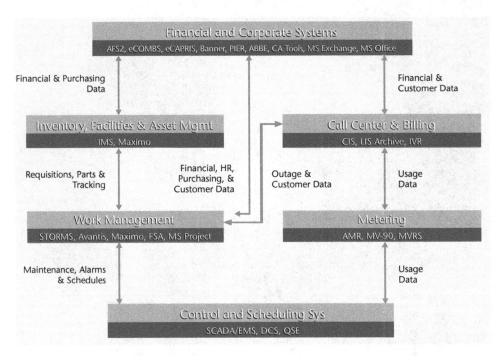

Figure 4.4 Austin Energy data flow diagram.

transparency, accountability and innovation we needed. In a sense, I achieved early success at Austin Energy by moving slowly, assessing and planning before prescribing and acting, all while gaining valuable information that contributed to ongoing project management as well as longer-term strategic planning. The technology governance structure enabled us to simplify project ranking and resource allocation and was truly the first and most important cultural change we achieved to enable the smart grid transformation at Austin Energy.

Technology Strategic Plan

In July 2004, the ITT Department released its first ever technology strategic plan internally. The plan included five key elements: (1) create and empower a technology governance structure (described earlier); (2) upgrade and standardize enterprise technology architecture (smart grid architecture); (3) implement project and resource portfolio management; (4) improve technology alignment company-wide; and (5) increase operational efficiency and quality company-wide. I would summarize our focus back then as shifting from operating a collection of business units to operating as a cohesive enterprise (which is also a critical aspect of advanced smart grids).

The most important technology goal of the plan was to build a smart grid. The steps to build the nation's first smart grid were driven by the need to simplify infrastructure, improve decision-making, adapt to faster changing business needs, improve disaster recovery and business continuity planning, improve regulatory compliance, increase quality standards, increase reliability, increase customer satisfaction, and reduce capital and operational costs.

The path to the smart grid ran directly through the technology governance methodology we used to ensure centralization of technology purchasing, decision-making, and business alignment, while remaining flexible and driven by business line executives and managers. These leaders acted as project sponsors, accountable to the enterprise for funding of the projects, business cases, ranking and alignment against the corporate strategic goals, and committed to delivering the benefits outlined in the business case that justified the investment.

Three key principles drove the executive support I needed for this transformation. First, we needed to *architect enterprise-wide, but deliver one discrete project at a time,* in order to demonstrate success, adoption, and culture change and build the necessary momentum for change. Second, we always emphasized that *perfection is the enemy of good*, where the search for perfect solutions risks forestalling good-enough solutions that would otherwise contribute to progress; we often settled for less than what we hoped for, but acknowledged each time that *progress and forward momentum were the keys to sustainable success*. Finally, we created a widespread understanding that building the smart grid is *a journey and not a destination*, a marathon rather than a sprint. This final principle was critical to maintaining morale, overcoming hurdles, and keeping the process underway.

The plan emphasized that achieving success would require *true top-down commitment to business process innovation and managing and rewarding culture change* that optimizes the attainment of higher levels of efficiencies and effectiveness and improved customer experiences. The plan would go on to highlight two key insights, namely that the *smart grid can be delivered sooner than most people think* and that *availability of technology is not the key constraint* to the delivery of a smart grid, rather, any holding back was generally based on *managing risk, selecting business models, and addressing internal political issues*. The plan also pointed out that the path to success would require a new way of thinking about our challenges as an industry and the solutions needed to empower a total transformation.

Getting the entire organization on the same page—instilling buy-in to the technology vision in addition to keeping the lights on—became the central tasks of 2004. A key element of the technology vision was to leverage common IP networking technologies to promote efficiencies and move the utility away from its traditional orientation around departmental silos, departmental applications, and specialty networks.

We already had installed a fiber backbone, and we had agreed on a strategy to expand fiber out to all of our distribution substations. Also in 2003, we had made the decision to deploy wireless networks to support our AMR and DR applications. However, our wireless networks were not integrated with each other or with the fiber network; and we had not adopted an integrated IP network vision by any stretch. In April 2004, I met John Cooper and began planning a project to promote IP networks internally, as detailed below.

The GENie Project: Considering an Integrated, Shared IP Network

Nothing summed up this new focus on an integrated energy ecosystem better than the GENie Project, which we began planning in May 2004. Over the next year, GENie proved a valuable tool to communicate the potential of a shared wired and wireless IP network and to promote buy-in of the enterprise focus of the technology strategic plan. We chose the name GENie, which stands for "geodesic energy network: information + electricity," because it captured the vision of a foundational IP network supporting new electricity architecture.

I hired John Cooper as the GENie project manager to raise the profile of IP networks inside Austin Energy. John spent the first several weeks assessing the situation, as I had done when I started back in early 2003. He then formed two cross-functional teams. The GENie Project Advisory Council was comprised of departmental management representatives oriented around communications or specific critical applications, providing coordination and guidance in drafting a strategy and implementing a trial. The Network Vendor Advisory Council, comprised of IP network communication and application vendors, served as subject matter experts and joint change agents to help communicate the IP network vision to the utility.

The GENie Project highlighted the challenge of migrating disparate Austin Energy networks supporting individual applications to common network architecture. The focus on a *single, integrated IP network* would be a sustaining aspect of the smart grid vision that would emerge over the coming years, but the first step was to support the business case for integrating a wireless IP network, which required a comprehensive review of current applications and networks and ultimately, selecting critical applications for a trial.

As shown in Figure 4.5, an integrated IP network can be the key to enabling a variety of applications, integrating relevant data, and providing significant program impact in a utility environment. The GENie Project identified an array of BPIs that would be enabled with a smart grid and set about to demonstrate the hypothesis of BPI and the efficacy of an integrated wired and wireless IP network with a trial. The GENie trial deployed a small Wi-Fi mesh network and ran early 2005 with notable successes and challenges. Looking back at GENie, we realize now that

Data + Applications + GENie Network

Business Process Improvement

App Pilots	More Data	Program Impact: Enhancement + Additions			
AMR	Interval Meter Reads Load Profiles Demand Usage Power Quality	Outage Detection from Manual Reads to AMR		Energy Management Services	
Conservation	Curtailment Status Chiller Flow Data Critical Peak Pricing	Demand Response	Load Shedding	Load as a Resource	TOU Rates
Customer Care	Outage Status Off-Cycle Reads Theft Detection	Outage Notification	Billing Consolidation	Trouble Acct Mgmt	Customer Loyalty Programs
Distribution/ Substation Automation	Outage Status Circuit Status Substation Power Quality	Mobile Data	Reliability Enhancement/ Outage Recovery	T&D Planning	Automated Controls
Security	Video Surveillance	Asset Protection		Homeland Security	

Figure 4.5   BPI and GENie.

we were ahead of the curve in 2005—much of what the current DOE American Recovery and Reinvestment Act of 2009 (ARRA) projects have done over the past three years, we envisioned in GENie, but back then, the technologies we used were less capable and more expensive than they are today.

The GENie project concluded in May 2005 with a report to the ITT department and Austin Energy executive team, which included among its many recommendations the deployment of a system-wide wireless IP network that would complement our fiber network and that could be shared among all Austin Energy departments. An important lesson learned at this juncture is that timing truly does matter. When the GENie recommendations were made, more specifically, when the principal recommendation to build out a system-wide wireless IP network was made, Austin Energy capital coffers were lighter than usual because of three mild summer seasons in a row (mild summers mean less air conditioning and lower cost utility bills). The recommendation to expand our wired IP networks with complementary wireless networking failed to win the approval of Austin Energy executive management.

On the other hand, the GENie report did serve its purpose of educating Austin Energy senior staff and departmental leadership inside the utility on the benefits of an enterprise-wide focus, continuing a process of paradigm resetting that I had begun with the technology infrastructure rationalization work since I had started at Austin Energy. Moreover, another key recommendation of the report, to upgrade the one-way AMR system into two-way AMI system, and to expand its coverage to match the entire service territory, proved more appealing and did receive executive approval. The incumbent vendor, Cellnet, now a division of Landis+Gyr/ Toshiba, proposed leveraging its underlying 900-MHz network two-way technology by deploying digital smart meters to all residential and commercial customer sites in Austin Energy's service territory. The upgrade and expansion from our initial deployment of 130,000 AMR meters began an ongoing series of projects that we would soon come to see as a major plank of our emerging smart grid program.

By mid 2005, the changes I had launched over two years before began to open up new opportunities and new levels of service. One of the key recommendations to come from the 2003 preliminary assessment had to do with the lack of visibility for Austin Energy customers on their bills. The ITT department recommended that specialized Web portals be developed to showcase energy information to multi dwelling-unit (MDU) managers. During the GENie project, we had spent time assessing the different applications that would be supported by a new enterprise-wide IP network. The potential to expand the partially completed AMR system, which at the time was deployed only to MDUs (i.e., apartments and condos) throughout the service territory, stood out.

It wasn't just about AMI though. The requirements of an expanding family of applications kept us busy planning and building strategy. Mobile workforce applications such as mobile laptops would benefit by pushing access to complex system data out to the edge, enabling field workers with new potential for enhanced productivity. DR also held great potential, given the significant consumer adoption and widespread success of the free thermostat program at that time, which communicated via a digital paging network. However, while the GENie Project had been successful in highlighting the potential of an enterprise-wide IP network, the departmental approach of single-purpose networks to support silo applications still had significant momentum and the vision of a unified, integrated wired and wireless IP network remained that—an aspirational vision.

Project Management

The work of the ITT department is conducted through programs and projects, making the establishment and institutionalization of an effective PMO a critical step on the way to a rationalized technology infrastructure. Without effective project management, it becomes routine for deadlines to slip, budgets to pass limits, and scope creep to increase the ongoing project requirements. Project tasks slip past their deadlines, and completion dates disappear over the horizon. We needed a PMO—and fast.

When I started, we had two people on the ITT staff certified by the Project Management Institute (PMI) [2]. PMI certification is widely regarded as a critical measure of quality, so one of the first things we did was send staff members to the PMI to get certification (today the utility has over 50 employees who are PMI-certified). We followed PMI standards and created an online PMO that allowed us to create easily accessible templates, reports, and dashboards, representing such project metrics as "customer usage" and "projects completed," as well as monthly steering committee reports.

In 2009, of over 4,000 companies analyzed by Gartner regarding project management maturity, only a handful worldwide had achieved Level 4 maturity. I'm proud that Austin Energy's ITT PMO was among that elite group.

In PMI process flow, as detailed below, there are five major states or "gates" whereby projects can be tracked: selection, initiation, planning, execution, and closing, with additional gates within each of those steps. Project progress is tracked in a PMO using the project state, its position on the project schedule, and the project complexity. Project criteria qualify projects as either "Run" (i.e., a project about keeping the light on), "Grow" (i.e., a significant enhancement to an existing solution), or "Transform" (i.e., a project to replace an existing solution with a new one).

The PMO allowed us to go from not tracking hours or effort at all when I arrived, to tracking every hour and every dollar, matching those to every requirement for every customer and every project to deliver the defined benefits. When I arrived, we had a ratio of 90% Run and Grow projects to keep the lights on, and only 10% Transform. By 2009, we had moved that ratio to 60% Run/Grow and 40% Transform. In other words, our standardization, process, control, and best-in-class practices had freed up an additional 30% of resources to apply to Transform projects.

Service-Oriented Architecture

In 2005, the ITT department also took the first steps on another significant project, the transition to a SOA. An SOA approach provides innumerable benefits over the long term for an IT department and the organizations it supports. I decided to power Austin Energy's smart grid using an SOA that would follow the principle of delivering presentation, process, and information as services to all stakeholders (including central power plants, distributed energy plants, the wholesale energy system, the transmission and distribution grid, the meters, smart appliances at customer sites, electric transportation, and the delivery of timely information via portals to all customer types).

In May 2006, we went live with the SOA-generated desktop application for call center representatives, integrating the billing system with the OMS. By allowing a function to appear once but be usable for any application needing it, the SOA eliminated extra steps in call processing, such as ensuring the completion of a customer validation within the OMS. Such validation is accomplished within the billing system, the work management system, and the financial system. Historically, network architectures would provide that every application must have all these services wrapped within the application. In a service-oriented world, however, the first step is to create one customer verification service, then make that service available to any application that needs the service.

For example, consider these steps in the OMS. First, a customer calls to report a power outage, and then the application verifies the customer's location and extracts the customer's status from a database. As soon as the call center agent transfers the information to a work order, the information travels to the OMS, which dispatches a service truck. Where this entire process originally took about five minutes to complete, with repetitive steps over multiple systems, the new approach took that time down to 1.5 minutes on average, a reduction of about 70%. The original process took five minutes because an employee would need to take such manual steps as looking up different items in different databases and applications, before he or she could eventually push a button to proceed to the next step. The cost savings were significant, but the improvement in customer service experience was even more dramatic. Before the SOA implementation, the application could handle only 4,000 calls per hour and required considerable waiting time online for customers. With SOA in place, on the other hand, the application could handle as many as 50,000 calls per hour. Similar if less dramatic efficiencies were also found in the back office.

Where traditional applications required programmers to install many of the same functions for each process in an application, using an SOA shrinks the overall amount of code used, standardizes functionality, and minimizes mistakes as employees use the same data sets.

The SOA effort inevitably drives improvements in a network, computer systems, data storage, data schema, and business process architecture layers. Doing one system and delaying others costs more in the end and takes longer. The SOA design process starts by mapping the current leading business processes, which provides valuable insights on the status of the enterprise. A session to set ambitious stretch goals follows it. Then, gaps are identified and a map is devised to proceed from the current to the future state, with quick win projects identified and given priority. I found it important to stay focused and work to the plan, and document lessons learned along the way. Finally, to maintain morale and generate momentum, we celebrated wins regularly and highlighted milestones throughout the organization as they were achieved, so that the entire organization understood the progress and could participate in the transition.

These three steps—concluding the GENie Project, adopting a system-wide AMI network, and installing SOA architecture—provided the required foundation for the future deployment of a smart grid at Austin Energy. The GENie Project helped to instill *a shared vision among utility* leadership regarding a *more holistic enterprise* approach and the potential to *leverage IP networking* to gain efficiencies in utility applications; the AMI network provided an opportunity to take a *bold first step* on that journey, and the SOA architecture provided sustaining benefits of *long-term cost savings* and *enhanced departmental functionality*.

Standards and Quality

Standards and quality are vital components when your goal is to become "the utility of the future." This vision, pursued by the ITT department as well as the entire organization, was stimulated and accelerated by our devotion to standards and quality.

The ISO provides a template for organizations with its ISO family of standards on quality management, which are designed to help organizations ensure that they meet the needs of internal and external stakeholders. From the outset, we established ISO certification as an organization goal, a unique objective in the utility industry. By 2007, we had reached one of several goals, becoming the first electric utility in the United States to achieve ISO 9000 certification [3]. This was an important milestone for many reasons, but one reason stands out in particular. Achieving this recognition became one in a series of external validations of our progress and success in our transformation. External validation proved critical to change the organizational culture and generate interest toward continuous improvement and sustained momentum.

Similar to ISO, the Information Technology Infrastructure Library (ITIL) is a set of concepts and practices for IT professionals to manage and run their operations better, providing detailed descriptions of IT practices and comprehensive checklists, tasks, and procedures that can be tailored to meet the specific needs of an organization. We chose to use ITIL to promote a deliberate transition to a more organized and effective department. Service support and delivery, infrastructure management, security management, business management—all these capabilities and more benefited from our use of ITIL. Each year, we hired external consultants to evaluate our progress on the ITIL continuum, measuring how we had matured as an IT organization according to the ITIL maturity model. Again, such benchmarking and external validation proved to be key tools to motivate continued progress among the ITT staff, but also to bear witness of our progress to the other departments inside Austin Energy.

The On-Line Service Catalogue is a great example of a tool that was inspired by ITIL; we published the catalog for our internal customers, starting with a hardcopy version issued in December 2005. Soon thereafter we were able to showcase with greater ease online the different service categories the ITT department provided within the utility. The catalog provided an overview of departmental mission and vision and an organizational chart and showed how the ITT department interfaced with other departments through the variety of governance mechanisms described above. The online catalog helped to move the utility along in our transformation to a paperless, computer-based organization, described in more detail in the creation of our digital platforms below.

Digital Platforms and Data Access

At the beginning of 2006, I became more focused on making the applications we had been creating more user-friendly and more effective as tools that would promote better management practices, such as management by objectives (MBO) using real-time performance measurements and feedback. For instance, we launched a project at that time that was focused on building user dashboards to leverage return on investment (ROI) tools, an internal project I pitched to leverage all the data that would be coming in from the different devices and applications we were deploying.

At the time, much of the data reporting was accomplished using paper reports populated with traditional data tables using month-old data, which required significant effort to produce, analyze, and interpret. To improve on executive data presentation, we would need to better understand data flows and formulate a data strategy. Tracking performance to objectives required the creation of KPIs that would provide the data hooks for user dashboards. We were able to develop dashboards that became standard for managers throughout the utility, providing access to critical performance data in a more accessible format on a far timelier basis.

Snapshot 2007

While the ITT department continued to promote a cohesive ecosystem approach based on the technology architecture vision, the cultural adherence to organizational silos remained a challenge. The inertia behind such organizational segmentation in an electric utility should not be underestimated. Remember, while ITT recommended a universal wired and wireless IP network to overlay the grid at the conclusion of the GENie Project, capital constraints and executive sentiment directed the program more along the lines of incremental additions of applications like AMI expansion, with distinct ROI objectives and project plans that conformed more closely with the business plans of the silos.

By the end of 2006, the ITT department saw the year's highlights divided into three major categories, all related to making more data available for better decision-making and lower costs. First, we upgraded technology infrastructure continuously, from the network out in the field to the back office. Second, we shifted more and more to digital platforms, with emphasis on Web portals for remote access. Finally, we leveraged network technologies and infrastructure to provide greater remote access for field operations. In this way, Austin Energy's smart grid began to emerge through a variety of incremental improvements, all oriented around leveraging digital technologies and improved network access to achieve Austin Energy's explicit mission to become the utility of the future.

The Smart Grid Emerges as a Tangible, Explicit Utility Theme: 2007–2008

By 2007, we were completing many of our initiatives and while we had made considerable progress, we still had a long way to go. Roger Duncan's vision of a utility more reliant on DER became more prominent with continued program success, particularly with regard to the nascent campaign to promote plug-in hybrid EVs. Smart thermostats had proliferated to such a degree that the utility had 100 MW of curtailable load when needed. In keeping with the digitization of our processes, we focused on completing the integration of an electronic bill payment and presentment system in 2007.

Smart Grid Infrastructure and End Device Integration

The emerging smart grid at Austin Energy in 2007 still resembled a collection of programs, processes, systems, applications, and sensor devices, more than a truly integrated ecosystem. However, if you looked closely even back then, you could see an interconnected smart grid ecosystem emerging, consisting of the mapping of processes, data, schemas, field applications, head end equipment, data center equipment, back office applications, and networks. All this activity was supported and orchestrated according to smart grid architecture we had laid out years before, so critical to our success.

First, a backbone of fiber assets connected all substations by 2006, with subsequent incremental additions. An assortment of wireless networks covered much of the service territory supporting various applications. In 2007, we also ran a BPL pilot, which confirmed what we expected—*BPL worked, but it was too expensive.* In 2006, we had also deployed a small downtown MetroMesh Wi-Fi project to support the World Congress of IT (an international conference Austin hosted that year). The Wi-Fi mesh project remained as a city asset in downtown Austin, which Austin Energy supported.

Beyond networks, the emerging smart grid in Austin was comprised of head end equipment, principally AMI smart meters (over 410,000 when fully deployed) and smart thermostats (ultimately over 100,000), but also including a small but growing number of smart devices attached to substation and DA gear. We also began planning in earnest to meet Roger Duncan's vision, anticipating the integration of connected EVs, ES, and solar PV systems, as well as HEMS devices that would emerge over the coming years in what we now call the connected home. In 2006, we had launched our Web site to promote the plug-in hybrid electric campaign nationwide and interest continued to grow in PHEVs, with retrofit kits brought in to help us start learning more about PHEVs. We also upgraded the Green Building Web site and launched an e-newsletter. Finally, in 2006, we had taken over the energy planning of the city's traffic lights, continuing in 2007 with incremental replacements of incandescent light bulbs by LEDs connected to an intelligent control system.

Completing the use of IT in the field were field mobile data applications, such as mobile mapping software and devices like laptops and later, smartphones that required connectivity and support for an emerging mobile workforce. A key challenge to providing data to field-based employees was the large bandwidth required for maps in service orders and trouble tickets. Consequently, initial applications required the workers to load high-bandwidth data on their devices before heading out to the field for the day.

Moving from the field to the back office, our SOA deployment continued, and we added asset management software, notably the Maximo software implementation that year. In 2007, we began a significant effort to build and integrate a data warehouse business intelligence program and related data centers. In our corporate facilities, we replaced our old PBX telecom system with new technology by Avaya, complemented by Cisco technology for our transition to voice over IP (VOIP) inside all Austin Energy facilities.

Beyond the ITIL compliance discussed earlier, we continued with BPI, change management, and communication initiatives. As a noteworthy aside, we had good success gaining organization commitment through the use of operational level agreements (OLAs), similar to SLAs that vendors commonly use with their customers to define the service they provide, but in this case, ours were internal agreements within the different technology groups. We upgraded our customer portals for MDU customers, and in our external communications, we began to articulate the utility goals of carbon reduction and clean energy, highlighting our utility of the future vision, where electricity demand and supply would be equally managed and accounted for.

Snapshot 2008

By 2008, after I had been at Austin Energy for five years, we could look back on significant success. The start of that year marked a landmark that we saw as the completion of our "four-year transition." The 2008 ITT Strategic Plan, which forecast activity from 2008 to 2011, included a graphic of "Austin's smart grid," which was aspirational at that time—we hadn't yet completed all our work, but by then, according to our organizational vision, we were indeed building a smart grid energy ecosystem. The network portion had been completed by 2008, and the infrastructure and key applications to enable the smart grid had been procured and deployed (e.g., data centers and major systems and applications,). The missing core element to complete our smart grid was integration, which would be the bulk of our work going forward.

By 2008, we had established a strong brand for Austin Energy with the utility of the future vision and our considerable progress to deploy a smart grid. As a city-owned utility, we not only were intent on leveraging the best practices and best technologies available, but we also felt a responsibility to stay involved with a variety of leadership groups nationally. Back then, I was a frequent speaker nationally (and occasionally, internationally) describing our progress at Austin Energy. Among the groups to which we contributed regularly and that we used to stay abreast of developments was Texas-based CCET [4], as well as such national groups as the Large Public Power Council (LPPC) CIO Task Force [5], IDC's Energy Executive Council [6], the Utilities Telecom Council [7], the Grid Wise Alliance [8], NIST [9], the DOE, [10], and the Federal Communications Commission (FCC) [11]. We contributed to the discussion where we could and gathered immeasurable help from our colleagues in these and other organizations, highlighting another key point: *We are truly on this journey together, and we all benefit from sharing where we can.*

As with ISO and ITIL, involvement with these organizations and the seven national technology awards we won over these four years (i.e., *Computer-world* Top 12 Green IT Company, *InformationWeek* 500, *HITEC* Top 100, *CIO Magazine* CIO 100 Award, *Computerworld* Premier 100 IT Leader, *Computerworld* Best in Class of Premier 100, and the Association of Information Technology Professionals IT Executive of the Year) provided the vital external validation that proved so important for us when conducting such a long-term program of complex megaprojects.

During our telecommunications and SOA transformation, our short-term goals in the ITT department evolved, and to accommodate our progress, we had to recast them as we went along. However, our mission remained steady: to deliver the first fully integrated and self-healing electric utility in the United States. The progressive approach we took with the technology infrastructure and the recognition we began to receive at this point were on par with the accolades other Austin Energy departments were receiving at that time for their own progressive achievements in energy conservation and incorporation of green energy into the resource portfolio.

In five short years, we had transformed ITT from a challenge to a strength, and in so doing we prepared the utility to leverage the investments we had made in applications and systems to improve substation automation, DA, metering automation, energy scheduling and trading, customer service, decision support, vendor management, and mobile field service, among other achievements.

Since the start of the decade, most of our applications and systems had needed fundamental upgrades or changes to keep up with technological advances and the growing requirements placed upon the ITT department. However, while we continued to upgrade key applications and systems, we had completed much of our foundational work by 2008 and could shift our focus to fulfilling our mission of building an end-to-end smart grid (extending from the central power plant and distribution system in-front of the meter to end consumer devices and DER located behind the meter).

And it's not like we were trying to hit a stationary target either—we would also need to achieve compliance with new FERC and NERC regulations on physical security and cybersecurity and meet ERCOT's redesign of the Texas' wholesale electric market (a transition from a zonal to a nodal market) [12] as well as upgrade to a new digital billing system, expand our mobile workforce management and dispatch capabilities, among other tasks.

By 2008, we could see that our transformation had allowed us to consistently eliminate complexity and offer new services, increased our service quality to both internal and external customers, increased transparency and accountability of both technology and business lines, eliminated all those single points of failure we had identified back in 2003, replaced legacy applications with state-of-the-art new technology, and increased trust throughout the ecosystem, as evident from the year-over-year increase in customer and employee satisfaction scores that both Austin Energy and the ITT department had achieved in the previous four years in annual surveys.

Having achieved much of what we set out to accomplish, the Austin Energy executive team created a new set of recommendations for 2008 going forward, which I've grouped into two major categories:

1. Adapt to external environment:
   a. Drive and support ERCOT nodal market enablement.
   b. Implement enterprise cybersecurity based on FERC/NERC and industry standards and best practices.
2. Implement, simplify, and expand:

a. Implement company-wide smart grid architecture.

b. Simplify and standardize applications into an N-tier framework.

c. Expand wireless network services.

d. Deliver business intelligence platform/framework.

e. Implement BPI (RUP, SDL, collaboration tools, and so forth).

f. Deploy B2E, B2C, and B2B portals.

g. Improve resource transparency and management.

h. Mature ITT's quality initiatives (ISO/ITIL/CMMI).

New Goals

These recommendations were closely tied to our larger, overarching goals for building a smart grid at Austin Energy. Most people now view a smart grid as a technology project, when in fact the smart grid is a means to a much larger end. At Austin Energy, we discovered that the smart grid was actually a journey comprised of multiple projects to achieve our corporate strategic goals. More broadly, the smart grid is a means for a utility to achieve its larger strategic goals, which for Austin Energy included the list outlined below.

First, *financial integrity* tops the list, including overall reduction of capital and operating expenses, which could be rephrased as, "being good stewards of the investments made by the customers, who pay for the utility in rates" and for investor-owned utilities, "being good stewards of shareholder investments." Our 2010 financial goal was to achieve AA bond rating.

Second on the list is *customer excellence*, which includes engaging customers in a new relationship with the energy they consume, and increasingly, the energy they produce, leading them from being passive and ignorant about electricity, to growing active and aware, and finally to becoming responsive and committed partners in energy consumption and production. Our 2010 customer satisfaction goal was to achieve 83% on the Customer Satisfaction Index of the JD Power Ratings for Utilities.

Third, *reliability excellence* is captured in such utility metrics as SAIDI and SAIFI. Our 2010 reliability goal was to achieve 40 minutes for SAIDI, and 0.5 interruptions for SAIFI. (From 2003 to 2008 Austin Energy managed to achieve the lowest SAIDI in the nation, which was very gratifying for all of us.)

The fourth goal is to significantly *reduce the carbon footprint*, as the production and distribution of electricity are principal contributors to CO_2 emissions worldwide. Our carbon reduction goal was successfully captured by a reached goal to become carbon-neutral in our fleet by 2012, but also a 2020 goal to attain 800 MW of energy efficiency (26% of our total generation capacity, which would allow us to avoid building more power plants).

Our final goal is *integration of renewable energy*, which accepts that the smart grid is an enabler of the transformation of the grid from a system based on burning fossil fuels to one that produces power based on technology and natural energy sources. Our renewable energy goal for 2020 was to reach a renewable portfolio standard of 35%, including 200 MW of solar energy generation (6% of total generation capacity).

After five years on the job and four years in a serious transformation program, working with all my colleagues I had accomplished the implementation of foundational projects and rationalization of processes according to our shared goals. The ITT infrastructure began to more closely resemble a cohesive ecosystem than a collection of systems and processes. With a sound foundation based on SOA and technology governance mechanisms, the ITT department could truly begin to provide the rest of the utility, specifically the OT managers, the required leadership and the data access and management tools they needed to reach our organization goals and realize our utility of the future vision.

Steps to Integrate an Energy Ecosystem

To achieve the vision of a smart grid, or more specifically our goals and our vision at Austin Energy, it became critical to integrate the various components and systems of the smart grid architecture so that the energy ecosystem would operate as a unified mechanism, not unlike the way that a variety of high-quality musical instruments in the hands of accomplished musicians make beautiful music in a symphony, according to the notes in the pages of the music and the guidance of the conductor's wand. Our work in building our own orchestra was laid out in our 2008 strategic plan, which featured the following five key strategic initiatives. These initiatives serve as our guide for this section, where I describe the final steps to integrate our energy ecosystem. They constitute a planning tool for anyone who would seek to do as we did—integrate technology systems to make a smart grid.

1. Create and Empower Technology Governance Structure

"Technology should not drive projects; technology is not in the driver's seat of this bus." This was the credo driving our technology governance structure. Going back to the smart grid architecture section earlier in this chapter, remember that the deliberate design of smart grid architecture starts top-down with the customer use cases and moves on to the processes needed to achieve those objectives, then on to the applications, data, and infrastructure needed to support those processes. In contrast, execution of the smart grid design starts bottom-up, with the infrastructure, moving on to data and applications, then to the processes needed to meet the desired objectives regarding customer impacts.

Starting with customer impacts was key to achieving the customer satisfaction goals described above, which drove the creation of customer use cases and objectives including this partial list: (1) managing outage restoration and notification times; (2) receiving usage information to better understand and manage customer bills; (3) being able to participate in energy efficiency and DR programs; (4) improving timeliness and accuracy of bills; (5) promoting turn-on and turn-off services; (6) enabling customers to manage smart home appliances via the Web or a separate display; (7) being able to participate in variable pricing programs; and (8) selling excess energy back to the grid.

In 2008, we went back to the technology governance plan we had instituted back in 2003, revising it to include an enterprise data council, which defined and managed the policies, procedures, standards, and guidelines for the creation, use, and management of data across the enterprise. As they had since 2003, the different groups within the technology governance plan proved instrumental in maintaining our focus on our objectives and coordinating activities within the enterprise.

2. Upgrade and Standardize Smart Grid Architecture

As I've detailed before, the smart grid architecture plan was the critical piece to drive our smart grid transformation. The biggest thing to emphasize here is that we were rebuilding an airplane while it was in the air—no small feat when you think about it. This is a principal challenge to utilities around the world: to transform their fundamental architecture even as they use their system to maintain reliability and keep the lights on at every moment. To accomplish this difficult task, we first created a parallel track.

Another way to look at our challenging task is to consider what highway departments go through when they build a new highway. We had to build a series of new superhighways right next to old highways, but with more lanes and far greater capabilities. Our focus was on redesigning the network, servers, data storage, processes, and controls, while upgrading applications and facilities. Successful transition required significant planning in advance, a large number of system integration projects, and a balancing of transition with cost management; we always had to keep financial impacts in mind to stay within budgets. We designed an upgrade path that went system-by-system, where we created a new system, then transitioned operations to the new system, taking time to test and confirm the transition before moving on to the next system and repeating the process. In this way, step-by-step, we evolved our smart grid architecture to a new, more modern and efficient design, while maintaining reliability and continuity of operations.

3. Implement Project and Resource Portfolio Management

The key to running a successful organization, and more specifically, transforming an organization, is to communicate the vision, then translate that vision into achievable steps and coordinate the activities of a variety of different organizations and individuals with widely divergent perspectives, according to a rational timetable and set of detailed instructions, while managing to a budget. We initiated such a rational approach back in 2003, and we began to enjoy the fruit of our early efforts later, as we saw the organization begin to transform along the lines of rational planning and execution.

We created processes according to guidelines from the PMI and implemented them with the help of our online tool, ITT PMO Live. The Web site provided real-time tracking of the myriad projects we were running, which helped us to manage our processes, projects, and documentation efficiently. A key tool in ITT PMO Live was the PMO workflow, which became our engine to evolve ITT functions and keep the rest of the organization informed and in-tune with the changes we were implementing.

Projects followed six steps: select, initiate, plan, execute, control, and close, which took a project from beginning to end. It's important to note that these steps focused on the relationship between the line of businesses (sponsors and customers) and the technology teams. Back and forth, changes were implemented and the tool was updated. We needed to track such things as project definition, requirements gathering, schedule, resources, alignment, justification, prioritization, and funding—for all of our many projects. The benefit of an online tool like this may seem self-evident, but let me emphasize anyway that it was this transparency and communication of the details that gave the organization the confidence it needed to go through with this transformation and ensure that all the day-to-day operations experienced minimum to no impact from the changes we were making. A key output of this tool was the summary report that allowed the entire organization to enjoy a shared understanding of the transformation as it was taking place.

Another key aspect of managing the transformation of a complex organization like Austin Energy was managing to a budget while achieving a large set of objectives, balancing business objectives, corporate goals, and technology realities and constraints. Portfolio management became the tool by which we managed to priorities, balancing processes, practices, and specific activities to perform continuous and consistent evaluation, prioritization, budgeting, and final selection of investments. This approach allowed us to provide the greatest value and contribution to help Austin Energy achieve its strategic objectives while balancing sometimes competing organizational interests, issues, politics, and agendas. We were able to make trade-offs among competing investment opportunities based on rational evaluation of benefits, costs, and risks.

4. Improve Technology Alignment Company-Wide

Alignment. Let's pause for a second and consider what an important concept that is to an organizational transformation. When a school full of students needs to evacuate the building rapidly in a crisis, they get in lines, exit the building, assemble on the schoolyard, and count heads. In a fire drill, everyone has a role to play and frequent rehearsals ensure that alignment will be second nature if a crisis ever occurs. A fire drill is a good way to explain the alignment methodology we took at Austin Energy to accomplish our objectives, but we weren't rehearsing for an event that might never occur. We underwent frequent fire drills, where we reviewed and accomplished successive steps to stay in alignment as we transformed the organization into a utility of the future via the implementation of a smart grid.

We looked at our transformation plans and recognized that we had two levels of alignment in order to implement the necessary technology solutions at Austin Energy, starting with our corporate strategic goals. First, alignment at the company-wide level was needed to optimize how we used our finite resources and minimized costs along the way—we had to remain efficient as we spent the public funds that kept us going. Second, we needed to align at the business unit level to maintain operational efficiency, human productivity, and meet customer needs: We needed to keep our business running as we transformed it into a new type of organization.

First, we needed to align the corporate goals with team and individual efforts. To achieve that, we chose what is called a *waterfall method*. Like a real-world waterfall, we envisioned repetitive processes that flowed downhill on paper based on successive completion of a series of tasks, just like a real waterfall flows down a cliff, pushed along by gravity. We captured team objectives and goals and action plans for working groups and individuals into planning charts that guided our activities. Detailed planning allowed us to marry individual accountability in tasks, actions, and goals to ensure collaboration with other individual tasks, actions and goals to achieve our organizational goals.

Second, we needed to align business unit goals and strategy with technology resources and plans. So we chose an *agile method*. As we aligned our technology objectives with business unit goals, we leveraged the tools we had created earlier, such as our principal tool for collaboration, the ITT/BU steering committee. In those meetings, the business unit leaders met regularly with ITT staff, which helped them make technology decisions in alignment with both their business unit strategy and vision and the smart grid architecture and our principal goals of *"cheaper, faster, and/or better—you can only pick two."* In this way, ITT became a trusted advisor as we replaced legacy systems with new technologies, in a virtuous circle of short-term and long-term successful collaboration.

Our agile method had at its core a step-by-step infinite loop with conditions where we systematically made old products and systems obsolete, created new projects, and mapped them to our architecture, operations, and customer requirements, always aligned to our goals. If we hadn't taken the time, if we hadn't dedicated ourselves to spending hours and days meeting to define how we needed to do what we did to meet our objectives, we would have wound up following the multiple processes and methodologies unique to the different departments. We would have run into *irresolvable conflicts* that would have slowed our progress and stymied our efforts to reach our shared objectives.

5. Increase Operational Efficiency and Quality Company-Wide

This section touches on one of the biggest challenges of implementing a smart grid vision. How can an organization balance the daily needs of running an operation with the vision of executing a strategy to transform the organization, all while staying within the boundaries and constraints of limited budgets? Today, many rate cases around the country contemplate raising rates to finance their smart grid plans. We didn't have that luxury at Austin Energy at the time, although we should have. Nevertheless and given our strategic goals, we crafted a plan to execute an evolutionary path to upgrade our infrastructure, data, applications, and processes according to our smart grid architecture, on a pay-as-you-go basis. Much of our financing necessarily came from savings from new efficiencies and optimization. We did devote new budgets to our upgrades, but they were relatively small compared to some numbers I see in rate cases around the country today. We looked for creative ways to shift budgets to the new paradigm, thereby freeing up capital to finance new projects and keep the transformation on schedule with current and reasonable new budgets. For example, recognizing that we were spending around $4 million each year for new meters to accommodate system growth and address meter replacements, we began to buy the new meter technology instead, even though we were out ahead of our AMI transition and network deployment.

I challenged the ITT department with the goal of *running IT like a business*. I started speaking on this topic in 2004, giving innumerable speeches where I described the process of taking our ITT department to ever higher levels of maturity in its evolution model, with the last stop being that mantra, "running IT like a business." In this new paradigm, the organization maintains a clear focus on controlling its resources and customers, measuring outcomes and learning from its mistakes and successes. *"Lessons Learned"* became a standard feature in our meetings, because we were educating ourselves at each step of our journey. We encouraged transparency in the delivery of services internally, in our change management execution, and in our optimization efforts, where we strove to achieve better yields at all times. We looked for ways to reduce our operating costs and studied what worked and what didn't, always stressing the importance of being 100% accountable. In this way, we worked to make the ITT department *an adaptive, fully optimized organization.*

The specific steps we took to achieve our strategic objectives while staying within tight budgets were guided by the concept of balancing supply and demand, which required that we get a handle on the demand for technology services through methodical, systematic control mechanisms with a finite supply of technology personnel and limited time. We did this by focusing on achieving operational efficiencies and improving quality practices, which took discipline, teamwork, and dedication. Given the dynamic nature of the environment and the shifting road map of a transforming organization, our ITT business unit was going to fluctuate with the needs of the multiple projects we ran over these transformative years.

To balance supply and demand, we restructured our organization to better manage relationships and resources. We needed to deftly manage expectations of business units that needed new technology resources to keep up with the increasing demands of their business lines; we did this by creating the new positions of *relationship manager* and *resource manager*, which worked in tandem.

One focused on listening to internal clients and building trust, becoming business line champions and the voice of the business line inside the ITT department, focused on outcomes from the project managers and business analysts. The other defined the boundaries of what was possible at any given time based on our finite resources, rationing the resource inputs to stay within well-defined budgets.

Over time, we formalized this approach with the PMO and a new customer relationship management organization. Along the way, we drew inspiration and guidance from ISO and ITIL methodologies and best practices. To focus on just one area, software development, we followed the capability maturity model developed by the Software Engineering Institute at Carnegie Mellon to assure a sophisticated and efficient software development life cycle. We adopted the rational unified process (RUP), a process standard that is widely recognized around the world.

The year 2009 saw the realization of many strategies as our smart grid came together. Inevitably, the world started to catch on to what was happening at Austin Energy and our smart grid journey to build the utility of the future. Our success had spilled over, and we were getting more awards that one could have expected. We received 11 awards in 2009 recognizing our smart grid leadership (e.g., UtiliQ #2 Company by *Intelligent Utility* magazine and IDC, *Computerworld* Top 12 Green IT Company, Energy Central CIO of the Year, *CIO Magazine* Hall of Fame Finalist, *Computerworld* Honors 21st Century Achievement Award, *Computerworld* Honors Finalist, *Computerworld* Honors Laureate, *InformationWeek* 500, UTC Apex Award Finalist, HITEC Top 100, and *Hispanic Business Magazine* Top 100 Most Influential).

As I close my reflections on our progress, I can't help but recall the challenges Austin Energy's GM Juan Garza outlined when I took this job back in 2003. He described an organization that lacked the funds to make the changes it needed to make. We found those funds by creatively changing the basic assumptions we were operating with, following an old adage I am fond of: *"When faced with an intractable problem with no good solutions, revisit your assumptions. Sometimes changing the problem itself leads to new solutions."* In our case, we addressed the way the organization operated. We changed our approaches to be more inclusive, transparent, and accountable and used such tools as creativity, standards, best practices, and external validation to drive the behavior changes that would release the funds we needed to help finance our transformation.

Lessons Learned

The case study presented in this chapter reveals a pathway to build a smart grid. Given the pioneering nature of this case study, we characterize this smart grid approach as SG1, recognizing that it offers a number of lessons learned, with two key lessons highlighted below.

Smart Grid Architecture Design Is a Necessary First Step

A principal lesson learned is that approaching the creation of a smart grid by incremental additions of applications is difficult, complex, and expensive. Stand-alone application choices with dedicated tools and networking resources cheat the utility from its future by failing to devise a complete blueprint for a smart grid architecture, which would follow these steps: (1) design the smart grid architecture first; (2) define the necessary use cases; and (3) review all the processes, selecting the applications needed to achieve the blueprint. That said, one has to start somewhere, and there were many debates along this journey concerning which system was at the core of the smart grid architecture. Choices included the GIS, the asset management system, the workforce management system, and the utility's control systems (SCADA/ EMS or DMS). After many what-if scenarios and debates, managers agreed that there was no one single way. In the end, the managers at Austin Energy chose the GIS as the core.

Leveraging Public Communication Networks Has Appeal If Necessary Conditions are Met.

Another important lesson is that utilities could leverage public communication networks to achieve their smart grid. At Austin Energy we met many times with public carriers to explore a partnership. They only needed to provide us with four things to dissuade us from building our own networks: (1) cost (fee for monthly access per device); (2) coverage (need to provide access to every required device throughout the service territory and with multiple networking technologies an option); (3) adequate SLAs (for priority access and restoration of service); and (4) commitment to deliver network access to every endpoint for a minimum of 10 years. These four elements proved difficult for public carriers to meet in the first decade of the new millennium, but as this chapter is written, several public carriers across the globe appear to be ready to meet the above requirements.

The vision of an advanced smart grid as a new, improved approach to building a smart grid flowed directly from the experiences at Austin Energy from 2003 to 2010. The transition was complicated by the mistakes we made and diversions we allowed along the way. For instance, we only recognized after the fact that our adherence to an applications-first methodology was a principal constraint to progress. As they say, however, "Hindsight is 20/20," and for us, hindsight revealed multiple insights. If I were to pick the single most important insight, however, I would capture it as follows.

> The considerable advantages in investing time and resources to get smart grid architecture in place, to get the organization behind the transformation effort, and to design and build (or lease) a foundational, integrated wired, and wireless IP network were keys to our success, by enabling much simpler and cost-effective deployment and integration of applications.

Highlights of the Austin Energy Smart Grid Journey

To recap, the following list of highlights represents the key lessons learned and achievements of the smart grid program at Austin Energy.

First, the *smart grid architecture* became the primary organizing principle for Austin Energy's smart grid transformation, which guided Austin Energy to deploy such components as portals, enterprise service bus, data warehouse, business intelligence, cybersecurity, project management tools, and fiber backbone.

Second, the upgrade of existing one-way wireless networks in 2007, and expansion of coverage to the entire service territory, enabled *full connectivity* to every device in the field, starting with the AMI system and its 410,000 smart meters.

Third, the deployment of *nodal market tools* to accommodate the changes at the ERCOT wholesale market included a new generation management system (GMS), a new network modeling system, an upgrade to the SCADA/EMS system, and, a new asset management, inventory management, and material management system.

Fourth, the deployment of *smart meters* throughout the service territory, reaching 100% coverage by the end of 2009, brought about edge-based intelligence and data gathering capability. The deployment of a new *meter data management system* (MDMS) complemented the AMI deployment to manage data and feed data into other systems across the company.

Fifth, the completion of the deployment of over 100,000 smart thermostats completed a *DR system* able to manage over 100 MW of interruptible capacity.

Sixth, the new state-of-the-art *billing system*, enabled real-time pricing and time-of-use pricing, as well as prepaid service, Web 2.0 services, mobile device access, sophisticated reporting and data analytics, and new services such as solar billing and EV billing.

Seventh and last, the deployment of a *DMS*, using many sensors across the distribution grid, integrated to the SCADA/EMS system, integrated with an upgraded GIS, complemented the existing AMI system and completed the first generation smart grid.

Beyond 2010, expansion of the smart grid involved deployment and management of new systems. The roadmap included smart appliances, smart EV charging stations, DG management systems, ES management systems, local area networks, and home area networks.

Envisioning and Designing SG2

Chapter 5 builds on this case study with the story of a groundbreaking community project that envisioned an energy Internet as a new approach to energy production, distribution, and consumption. The Pecan Street Project emerged as an idea while the smart grid at Austin Energy was still being deployed in September 2008, blossoming into a full-blown project with the help of Austin Energy and other community leaders starting in 2009. Chapter 5 describes the emergence of an advanced smart grid vision from both the lessons learned during the seven-year development and deployment of the nation's first utility-wide smart grid at Austin Energy and throughout the Pecan Street Project from 2009 to 2013.

Endnotes

[1]   www.theopengroup.org.

[2]   http://www.pmi.org/.

[3]   http://www.metricstream.com/pressNews/pressrelease_63.htm.

[4]   http://www.electrictechnologycenter.com/.

[5]   http://www.lppc.org/index.htm.

[6]   http://www.energy-insights.com/eec/index.jsp.

[7]   http://www.utc.org/.

[8]   http://www.gridwise.org/.

[9]   http://www.nist.gov/index.html.

[10]   http://www.energy.gov/.

[11]   http://www.fcc.gov/

[12]   ERCOT reorganized from alignment around a handful of large zones to a nodal approach with many more small nodes to provide market incentives, align transmission costs, and encourage power resources to be located closer to the load they serve. The wholesale market realignment in Texas required all utilities in Texas that were active with ERCOT to adapt.

CHAPTER 5

Envisioning and Designing Smart Grid 2.0

In Chapter 4, we thoroughly documented the significant progress achieved by Austin Energy over the past decade with its smart grid effort, as the utility's management and staff demonstrated national leadership in a variety of programs, from green power to EE to smart thermostats to its pioneer smart grid. With its Green Choice program, for instance, Austin Energy acted as the principal driver to kickstart the West Texas wind farm industry in the early years of the new millennium. Up to 2011, and for the *ninth year in a row*, Austin Energy was the number-one utility in sales of green power, when measured by total kilowatt-per-hour retail sales, according to NREL. Austin Energy dropped to second place in 2012.

However, there are risks in being a pioneer. Austin Energy's solar PV rebate program historically offered some of the most aggressive rebates nationwide, but its success as a pioneer led to a problem encountered mostly by mature rebate programs: Adjustments to the program were required after 2009. As the program grew ever more popular, it became oversubscribed and needed to be recalibrated to accommodate changes in federal tax incentives and the maturity of the local industry. The value of the solar program developed at Austin Energy and originally released in 2004 presents a template for solar PV that is being modeled at utilities across North America in 2013. The Power Partner smart thermostat program has so far provided over 120,000 residential and commercial customers (approximately 25% of all Austin Energy customers) with free digital thermostats in exchange for their permission to cycle air conditioners on and off during the critical peak periods of the hot summer months, amounting to over 110-MW capacity, making it one of the largest and most successful DR programs nationwide.

With a little help from its friends, Austin Energy started the Plug-In Partners program, which became a national movement that brought cities around the United States together with pledges to add plug-in hybrid EVs to their city fleets, providing major car manufacturers the confidence they needed to commit to new EV manufacturing goals. Finally, Austin Energy completed the replacement of its entire meter stock—410,000 meters— with new smart meters, making it the first utility in the nation to have an operational AMI network throughout its entire service territory integrated to the rest of the grid. However, as with the solar PV rebate program, there is a price to going ahead of the pack, as Austin Energy's smart meters are of an earlier generation and are not capable of communicating energy consumption data directly to in-home devices, which has become standard in the industry.

Introduction

Chapter 4 described the foundation of Austin Energy's original smart grid in a rational IT infrastructure, smart grid architecture, and an integrated IP network communications network. In this chapter, the trend lines outlined in Chapter 4 lead to the Pecan Street Project, a landmark research program that provided valuable insights on transitioning from an application-led SG1 to a network-led SG2. SG2, an integrated, advanced smart grid, will provide innumerable advantages over the conventional approach to smart grids, not the least of which will be lower TCO, more rapid deployment, and more flexibility to accommodate unexpected changes in the future.

This next stage in the smart grid journey reassesses the vision of SG1, where the goal remains the same—the transformation of the energy utility from its roots as a centralized, fossil fuel-driven analog power grid—but the question asked this time is different. As technology progresses and as energy users gain new awareness, integration of new elements with the smart grid will become ever more essential, suggesting that the smart grid should be designed from the start with robust future needs in mind, such as integration of edge resources, alternative infrastructures, and consumer activities and ideas. The Pecan Street Project, which we'll explore in great detail in this chapter, originated based on the belief that significant benefits are possible through integration—of DER systems, transportation, and water—on *a common, more resilient smart grid infrastructure.*

Few utilities so far have paused to look at the entire universe of strategic options; the Pecan Street Project in 2009 may well have been unique in its bottom-up, inclusive, and community-oriented approach to visioning and strategic planning. From the start, the Pecan Street Project charted new territory, expanding the smart grid focus well beyond DA and AMI, leveraging the significant experience embedded in previous efforts at Austin Energy to envision a plan to build a network that could fully integrate dynamic *DER*, including water and transportation infrastructure, to help the utility achieve sustainability.

This key insight on the importance of IP networking as a new approach to smart grid came as the idea of the Pecan Street Project was first floated in mid 2008. Austin Energy management embraced the opportunity to participate in a broad-based community effort to help it in its journey to reimagine the provisioning of energy services for the coming decades and to explore a network-driven smart grid transformation. The community leaders who launched what came to be called the Pecan Street Project envisioned clean energy as the foundation of new economic development in Austin that would also provide energy security and environmental health over the long term for Austin citizens. Austin Energy managers saw even more. Throughout the Pecan Street Project, Austin Energy managers gave their support in leadership and man-hours to make the Pecan Street Project a success but also drew valuable lessons from their interactions with industry, academia, non-profits, and community volunteers. Moreover, the effort paid off in helping Austin Energy determine a new future to match its significant early progress in creating its SG1, whose efforts positioned Austin Energy to learn even more from the Pecan Street Project, which came to represent a second bite at the apple, an opportunity to refine lessons learned and to craft a new, more effective approach to smart grid, which we have labeled SG2 or the advanced smart grid.

The Pecan Street Project: A New Approach to Electricity

In 2008, a group of Austin civic leaders began meeting to discuss an intriguing proposition: Could Austin repeat early economic development success in computers and semiconductors in the emerging field of clean energy? Three notable Austin high-tech successes in the 1980s and 1990s became nationwide models for economic development success. First, the Microelectronics & Computer Technology Corporation (MCC), formed in 1983, was a research consortium financed by 12 technology companies to promote research in supercomputers and related technologies. Second, the SEMATECH semiconductor research consortium launched in 1987 as part of a strategy to preserve national competitiveness and competency in semiconductor chip manufacturing, ultimately became a local economic development engine over the next two decades, ensuring Austin's position as a global center for semiconductor manufacturing. Furthermore, moving beyond semiconductors, SEMATECH would help burnish Austin's high-tech reputation, launching Austin as a mecca for local tech startups during the Internet boom that followed a decade later.

City leaders gathered in 2008 to ponder how the creation of a clean energy consortium could similarly attract established clean tech companies to Austin and help incubate clean energy start-ups. After all, in the emerging clean energy economy, regional economic growth had become inextricably linked to technology innovation, environmental health, and green job creation. By early 2009, Pecan Street Project had come to be about much more than economic development.

Many have asked, "Why the name 'Pecan Street'?" In the early nineteenth century, when city planners laid out a transportation grid for the new capital city of Austin on the north bank of the Colorado River, they named the north–south streets after the rivers in Texas, and the east–west streets after the trees of Texas. Halfway up from the river to the Capitol building ran the main east–west artery, named Pecan Street. In late 2008, when city leaders brainstormed a name for their new community clean energy project, they looked for something distinctly Austin and finally chose to name the project after the original Pecan Street, now widely recognized as Austin's Sixth Street, home to one of its burgeoning entertainment districts. Thus, a nineteenth-century transportation grid lent its name to a project that would provide key insights into a twenty-first-century energy and information grid.

The Pecan Street Project's inclusive, community-led approach reflects in many ways the central role that Austin Energy has played and continues to play in the life of the Austin community. Although Austin has grown into a leading metropolitan area for Texas and the nation, it has somehow managed to retain its small-town charm. Tackling smart grid planning in an inclusive, community fashion was very Austin-like. Volunteers from *large IT corporations* (Cisco, Intel, GE, Oracle, IBM, Freescale, Dell, Microsoft, Applied Materials) and *local small business innovators*

(Xtreme Power, Helio-Volt, Austin Technology Incubator, Greater Austin Chamber of Commerce), together with *world-class researchers* (University of Texas, SEMATECH), *environmentalists* (Environmental Defense Fund), and *one of the most progressive energy utilities* (Austin Energy) comprised the teams that made up the first phase of the Pecan Street Project. A key benefit of being so inclusive was to harness innovation and creativity in a friendly, open environment. Leaders hoped that the diverse mix of volunteers they assembled would produce fresh thinking and insights not only on timely, critical issues facing the industry, but also on thorny, old problems that have bedeviled the electric utility industry since its inception.

A New Design, Business Model, and Empowered Energy Consumer Class

The utility business model launched by such early electricity pioneers as Edison, Westinghouse, Tesla, and Insull in the late nineteenth century made electricity reliable and affordable and drove economic growth for over 100 years [1]. Now, however, the grid faces significant disruption, challenged by the need to reduce reliance on fossil fuels, but also by the rise of increasingly viable technology-based alternatives to grid power. As an early adopter community, Austin found itself at the forefront of this wave of change. Its citizens were eager to adopt the latest technologies and have naturally gravitated toward the edge of change. However, this early progress highlighted a conundrum now being recognized industry-wide: Microeconomic efficiency results in new macroeconomic challenges—every energy dollar a utility customer saves by applying new technologies becomes a revenue dollar lost for the citizen-owned energy utility. Technology integration is on the critical path to infrastructure reform, but so is the question of a new, business model that can more readily adapt to a steady evolution in technology-based energy solutions.

Community leaders in Austin realized that they had to approach such critical changes in deliberate fashion, since significant success with alternative energy would risk undermining the utility's financial foundation, even as it *increased* demands on services only the utility can provide into the foreseeable future, including construction and maintenance of transmission and distribution lines, backup power, and reliability. To compound the financial risk inherent when declining revenues meet increasing costs, the city has also come to rely on Austin Energy revenue transfers over the past decade to fund city services—as a city department, Austin Energy transfers roughly $100 million each year from its profits to Austin's general revenue fund. Without this transfer, budgets must be cut, or new revenues must be raised.

The original mission of the Pecan Street Project then was threefold. First, it sought to *reinvent the city's energy, water, and transportation systems* through integration of the most advanced technologies and systems, while maintaining financial and environmental sustainability. Second, it would foster the creation of *new clean energy industry companies and jobs* in Austin. Third, it planned to provide *a replicable model for systemic change* for other communities in the United States and around the world.

Other municipally owned utilities, investor-owned utilities, and cooperatives face a similar transitional challenge: Like Austin Energy, they will need *a new sustainable business model* that will support the transition to a twenty-first-century energy economy, making full use of advanced DER technologies while maintaining the infrastructure necessary to satisfy a growing demand for reliable electricity.

The twin challenges of DER integration and business model transformation combine to make what some social scientists have labeled a *wicked* problem [2]: a problem characterized by complexity and multiple subproblems that lacks clear solutions or right answers. Unlike lesser problems, wicked problems are not so much solved as they are whittled down into smaller bits to make them increasingly more manageable. Collaboration among a broad group of stakeholders enables iterative problem solving by way of pilots, flexible platforms, and experimentation that increases clarity and coherence over time.

Another way of looking at the Pecan Street Project, then, would be as an assessment of the challenges it faced during the transition to a new energy ecosystem: *technology integration* and *business model development* on the one hand—the wicked problem—combined with the third great challenge, *community engagement*—the solution to the wicked problem.

Starting with Strategy

Undertaking a study of an integrated complex system like an electric utility requires a strategy. The Pecan Street Project organizing committee met and determined their strategic approach would be to divide its objectives, tasks, and volunteers into logical groups, and then throughout the first six months of Phase One, adjust and recombine the groups as indicated by progress and discovery. From the long view, the planners also recognized that accomplishing the original goals would come in phases; this effort was Phase One.

The community-wide review and discussion of strategic issues regarding the creation of an energy Internet, Phase One, wound down after July 2009 when the teams quit meeting. The project planners acknowledged a need to ratify the *subjective* Phase One conclusions that we'll focus on in this chapter, with *objective* conclusions to be based on the more quantitative work that would follow in Phase Two. With a few adjustments, the steering committee from Phase One reconstituted itself as the board of directors of the new nonprofit Pecan Street Project, Inc., formed to pursue more quantitative analyses, pilots, and demonstration projects in Phase Two. Now in its fourth year as an innovative R&D facility with the simpler title of Pecan Street, Inc., Phase Two work began with a $10.4 million ARRA Demonstration Grant award from the U.S. DOE in November 2009. Remaining sections of this chapter focus on Phase One, including an update on Pecan Street, Inc.

Change on Three Dimensions

Phase One told three storylines, aligned with the challenges enumerated earlier. First, the story was about technological change, more specifically, the five essential components of DER (EE, DR, DG, electric transportation including EVs, and ES). The integration of water infrastructure and integration of the separate DER components into a cohesive whole called the smart grid completed the technology change storyline.

Second, the story was about the *impacts* of technological change and the *requirements* of such massive integration, leading to a discussion of options for redesigning the rate base and alternative sales of electricity services, collectively referred to as *business model transition.*

Finally, the story would need to consider critical societal elements and the *community engagement* needed to accomplish the requirements on the technology and business fronts.

Getting Organized

Drawing on both the local community and experts from academia and industry, the Pecan Street Project ultimately organized some 200 volunteers, experts in the fields of energy, telecommunications, software, hardware, project management, policy, finance, behavioral change, water, and sustainability (see Figure 5.1). The volunteers were placed into a dozen teams to brainstorm the broadest array of possibilities according to a common strategic outline. In Phase One, the 12 teams met weekly over a seven-month period and ultimately generated hundreds of discrete ideas. It soon became apparent to the steering committee that the value of the project would

Pecan Street Project Teams

1 Distributed Generation

2 Energy Efficiency

3 Demand Response

4 Transportation

5 Water

6 Energy Storage

7 Operations, Systems Integration and Systems Modeling (Smart Grid)

8 Business Model

9 Customer Interface

10 Regulatory and Legislative

11 Economic Development

12 Workforce Development

Figure 5.1 Project teams.

come from both the detailed assessment of different options in each of the team categories, and from a detailed assessment of DER integration using the connective fabric of the evolving smart grid. As the project progressed, the complexity of the undertaking began to unfold, given the increasing pace of innovation in a variety of clean energy segments, a variety of changes on the policy front, and the critical need to keep the grid functioning during any transition.

While nine of the 12 teams dove deeply into the particular arcane details of their six-month study plans, three teams were more concerned with integration of the project scope into a cohesive whole. Team 7, the Operations, Systems Integration and Systems Modeling Team, soon labeled "the Smart Grid Team," tackled the issue of technology integration, system redesign, and system modeling. Team 8, the Business Model Team, grappled with the financial and economic implications of the monumental changes facing the utility industry. Team 9, the Customer Engagement Team, focused on previous surveys conducted by Austin Energy and work at the national level to understand and integrate customer energy use and behavior into the new energy paradigm.

During the course of the seven-month process, each of these three teams had individual meetings with the more detail-focused teams. In June, as the process wound to a close and reports were being written, the 12 teams were reconstituted into four "super teams" focused on key DER issues—EE, DR, ,DG, and EV. Members from Teams 7, 8, and 9 worked more closely within the super teams to provide further integration and support for idea cohesion. Throughout the Pecan Street process, Team 7 remained the team with the greatest focus on system design and modeling, on the transformative technology and on the task of integration with a new more resilient version of the smart grid; what we now call SG2, or the advanced smart grid.

Operations, Systems Integration, and Systems Modeling: Team 7

Team 7 met about 30 times between February and August 2009, establishing a strong task orientation based on effective communication, shared workload, and healthy group interaction. With each successive weekly team meeting at Austin Energy offices in downtown Austin, Team 7 developed into an umbrella organization, pulling together the work of the other teams in the Pecan Street Project within its Pecan Street Architecture Framework, a tool developed in the first meetings. [3]. To create a comprehensive, integrated new electricity paradigm, Team 7 continuously refined the Pecan Street Architecture Framework and related systems, technologies, and integration points.

From 2003 to 2009, Austin Energy had evolved its operations and systems to build the nation's first smart grid in a major U.S. city. The Pecan Street Project offered Austin Energy the opportunity to explore alternatives for expanding on the original smart grid focus of internal IT system redesign and support of individual systems like DR and DA. The team was able to envision a more complete transition that would prepare the organization and the city for the system integration required by increasingly rapid advances in network and digital technology, as well as the disruption that was sure to follow. After all, the introduction of new technologies into a distribution grid, while transformative, can also be extremely disruptive, as the following forward-looking case study illustrates [4].

Use Case: Influx of EVs

In March 2022, when Austin hosts the 35th anniversary of the unique South-by-Southwest Music/Film/Interactive Conference, the potential for disruptive impact embodied in the thousands of EV owners who come to town as part of the hundreds of thousands of attendees will become reality. Without prior planning and programs in place, specific distribution feeders will risk being overloaded by the massive influx of large charging loads. What's more, energy bills for those whose outlets are used could be dramatic. Moreover, charging stations will be difficult or impossible to find in certain areas, while other charge stations could sit vacant and unused.

Arguably, no electric utility is prepared for such an influx of mobile, unpredictable load. However, thanks to its SG2 system and substantial forethought, Austin Energy manages to weather the storm brought on by the massive influx of EV enthusiasts. In advance, online, or logging into onboard navigation systems, drivers tap in their destination of "SXSW," and their computers, phones, or EVs are connected to Austin Energy's advanced smart grid, which begins the planning cycle with the user. On registration and account creation or reactivation, the smart grid obtains the ID information for the EV, noting such unique descriptors as battery type and capacity and charging capability. Ideal locations for charging and unique pricing arrangements are negotiated based on the user's schedule and profile preferences, as well as the utility's load parameters. Charge rates reflect not only the time of the day, but also grid congestion and location in proximity to common high load areas. Special rates are assigned for charging stations at the ends of the light rail system, for instance, to spread charging load into the residential districts. Load flow of the grid will drive charge rates at different stations, putting a premium on dense feeders near industrial and large commercial loads.

While EV loads promise to stress the grid in new ways as described in the use case above, the potential for disruption is not limited to EVs—the entire DER category deserves more detailed attention when it comes to the potential for system disruption. From the perspective of the Pecan Street Project and for our purposes, a closer look at the term *DER* reveals the following elements [5].

1. *EE* begins with the building infrastructure and the energy appliances included inside. EE may often be low-tech, but EE can also incorporate technology advances to seal the building envelope and replace old inefficient energy appliances with newer, more efficient models. *Residential* EE includes such advances as smart appliances, in-line water heaters and solar thermal technology, redesigned efficient windows, radiant barriers in attics, and spray foam insulation. *Commercial* EE includes similar technologies at the building and appliance level, as well as BEMSs and much more.

2. *DR* is facilitated by smart end devices/appliances located beyond the meter, which includes both HEMSs for residential DR and BEMS for commercial DR. DR may feature smart thermostats to manage the use of HVAC systems, smart lighting with new LED technologies, and set-and-forget user interfaces to automate DR.

3. *DG* includes much more than rooftop solar PV, which gets most popular attention. Other technologies in this category include microwind generation, microturbine CHP generators, geothermal systems, and any new invention that fits the mold of smaller, more numerous generation plants closely aligned with the load they are intended to serve, rather than providing power to be distributed long-distance over a grid.

4. *ES* from the Pecan Street Architecture Framework perspective was divided in two groups, mobile and fixed. *Electrochemical* storage (e.g., batteries) may be available as *mobile* ES (e.g., plug-in electric hybrids and pure-play EVs) and *fixed* ES, which may include utility-scale storage, community ES (CES), and premises or personal ES. *Thermal* storage can be either *cold* thermal ES (e.g., ice machines to make ice and supply chilled water to HVACs and chillers, or HVACs cycled during the day to prechill homes) or *hot* thermal ES (e.g., solar thermal devices that store heat in water and/ or melted salt or energy thermal systems (ETSs) that store heat in bricks to avoid peak energy consumption in cold climates).

The Pecan Street Project and Team 7 in particular intended to provide a road map for incorporation of all these aspects of DER into the utility's operations and systems to transform utility potential, while mitigating and managing any disruptive effects, both leveraging and evolving the smart grid.

Pecan Street Architecture Framework (PSAF) Design

Team 7 began its work by crafting a new framework, dubbed the PSAF, which would include new processes, system approaches, relationships, and technologies, thereby extending Austin Energy smart grid efforts to cover the new Pecan Street Project focus. Putting together a model for future integration proved a significant challenge, as it needed to be robust enough to incorporate multiple DER elements, as they are developed and deployed while managing transitional business model issues. Essentially, the integration model would need to constitute a dynamic plan for maintaining critical operations and infrastructure during a transition to an increasingly hard-to-predict future. The flexible, iterative model would need to outline a means to progress from an existing layer of technology—the established SG1 platform developed and deployed over the last several years—to a new platform that could be called SG2.

Key elements used to drive such planning in Phase One of the Pecan Street Project included not only last-mile communication and adoption rates of DER elements, but also development and adoption rates for ES, which provides such a dramatic transformation that it should be called out from the other DER elements. Phase Two analysis and demonstration projects would be used to test assumptions developed in Phase One to establish quick wins, identifying potential economies of scale and successful deployment methodologies. SG2 design needed to emphasize efficient investments, and when sufficient funding existed, recommended investment in longer-term programs. SG2 planning needed to accommodate a range of goals in the short, medium, and long term.

To understand the design of a future integrated energy delivery system, Team 7 started with a review of the traditional electric utility supply chain and composed a current state architectural framework (CSAF) for the utility environment. The framework is characterized first by *domains*, as follows: (1) *electricity generation* from turbines driven primarily by steam and to a lesser degree, water; (2) *wholesale energy operations*, providing for the provisioning of individual electricity distribution systems; (3) *electricity transmission* over long distances at high voltage levels; (4) *electricity distribution* in local areas at lower voltage levels; and (5) *retail energy operations* including metering, billing, and customer service.

These are the traditional five domains that make up the energy ecosystem from an operational and systems standpoint. A key challenge for the electric industry in the coming years and decades will be to adapt this fundamental supply chain to future requirements by adding DER as a sixth domain and to enhance the transmission/distribution grids by transforming them into advanced smart grids.

Central to the task of creating a new framework to remap these domains was a paradigm shift in the vision of an electricity distribution system. From the *traditional* perspective, the principal task of the network system has always been to distribute centrally produced power down the line in a one-way flow, out to the edge for consumption, where the electrons are used to accomplish a variety of work tasks by producing light and powering appliances. From the new *transformational* perspective, the principal task of the network system will be to distribute power produced from a much more diverse pool, not only from large remote centralized power plants and medium-sized intermediate power plants, but also from much more numerous, but much smaller, distributed power plants located out at the edge and designed to be near the load they serve. The addition of such edge power devices introduces the risk of two-way power flow, from the edge back along a distribution feeder when distributed production exceeds consumption.

Further, this new perspective also needed to incorporate two other elements that at the time were rarely incorporated in utility strategic planning: first, DR— on-demand curtailment of energy consumption—at a far greater level of integration with all types of customer levels, and second, both fixed and mobile ES, which dramatically changes how and when energy can flow and dramatically alters the economic equation so central to utility planning and operations. (In essence, the traditional paradigm has depended upon just-in-time energy production to match energy consumption.) In short, the new paradigm for energy distribution can be described as *two-way flow of both energy and information* in dramatically different ways than the original power distribution design ever encompassed.

In summary, the role of Team 7 was to extract necessary functionality from group discussions and to use homegrown integration and design tools (e.g., the self-developed system architecture matrix), as well as external tools like the final NIST Interoperability Smart Grid Roadmap to map the elements that make up each domain of the PSAF. By comparing current solutions with requirements developed during the analysis phase, the team identified gaps to fill in order to transition the utility to a new state and meet SG2 needs, most notably providing for effective DER integration.

Using the NIST document, for instance, a review of the different gap categories identified by Team 7 revealed the following strategic issues:

- *Adequate cybersecurity.* Current dedicated application networks (e.g., AMI network, DA network, and DR network) lack the capacity for multilevel/multilayer security from end-to-end devices, to network, to the utility network operations center. The elements of a robust security system for SG2 must include smart device identity, secure digital keys and certificates, secure authentication and encryption, secure communication, and secure data transmission. Such a system would leverage sophisticated hardware and software with multilevel/multilayer solutions, allow for discovery of the device at the software and hardware layer using public and private keys, and be capable of authentication and authorization via the network layer and verification by the utility control center, all with secure encryption and replay protection.

- *Bandwidth connectivity.* SG2 requires a future-proof communication network—an IP network that can provide connectivity throughout the service territory at speeds and capacities necessary to manage electricity in the new environment. Real-time data transfer capability will be needed for mission-critical control systems (from 20 to 100 milliseconds), to support emergency situations, as well as real-time resource management and dispatching. Moreover, the exponential increase in data traffic that can be expected based on the proliferation of devices throughout the utility's service territory must be accommodated (minimum capacity of 2 MB).

- *Dispatch scheduling.* Today, distribution utilities lack the engagement, connectivity, cybersecurity, and pricing rules needed to maintain system reliability and enable third-party DER owners to be significant participants in the new energy ecosystem. These new parties and new resources must be able to engage and disengage with the grid on a near automatic basis. The number of transactions inside a DER-enabled distribution grid will far exceed current capacities and not only drive the need for more communications bandwidth (i.e., IP networking), but also a new set of rules and standards for distribution grids modeled on those currently practiced in transmission grids nationwide (e.g., economic dispatch).

- *Standards and interoperability.* The current distribution grid operates under legacy standards and proprietary equipment, whereas SG2 will need to be able to incorporate a variety of vendors, applications, and hardware and ensure that they will all interoperate according to the old standards (DNP3, MUD-BUS), as well as the new standards (IEC 61968, IEC 61850, and IEEE 1547).

In close collaboration with the other Pecan Street Project teams, Team 7 expanded its vision beyond gap identification to include new processes, system approaches, and relationships, ranging from alternative forms of generation to smart grid distribution upgrades, the leveraging of smart meter functionality, and dramatic new potential on the demand side of the grid. At each step of this dynamic process, new technologies, systems, and integration points will facilitate efficiencies and enable cleaner production, distribution, and consumption of energy.

At this stage, it makes sense to pause and walk through the process in more detail, showing how the process unfolded chronologically and how tools emerged for the problems at hand. After all, the processes and tools Team 7 developed and used in Phase One provide a roadmap and model for other utilities with similar challenges, which may help managers explore and define paths for accomplishing the necessary changes to operations and systems and identify options for the utility and the community.

As stated previously, the team started by creating the PSAF with detailed discussion to determine how to define and organize the domains and subdomains for effective representation of the Austin Energy supply chain. First, the team used *mind-mapping software* to interactively diagram the domains and subdomains that comprise the local energy ecosystem and produce the PSAF. The new framework not only documented the current utility supply chain, from energy generation to consumption, but also provided a template to guide the integration of DER. The following section is an excerpt from the Team 7 notes from the day in February 2009 when the PSAF was laid out.

Power Engineering Concept Brief

The domains in the PSAF are listed below, with subdomains highlighted. One of the first tasks Team 7 took on was to map these domains using mind mapping software, as described above, to produce the PSAF.

Domain 1: Central Generation

Team 7 determined that Domain 1 would not be included in its project, instead focusing attention instead on more in-depth discussion and exploration of the remaining domains. Subdomains in this domain include: Subdomain 1: ERCOT Grid; Subdomain 2: Power Plants; and Subdomain 3: Wholesale Market.

Domain 2: Generation Market Operations

This domain features two subdomains: Subdomain 1 (Retail Market Operations) is defined as "the area under Austin Energy control," while Subdomain 2 (Resource and Generation Management) is defined as on/off and any other load control is the point at which other PSP teams will link in their work product (Teams 1 (DG), 3 (DR), 4 (EV), 6 (ES) linked to this subdomain).

Domain 3: System Operations

The four subdomains include: Subdomain 1: Transmission; Subdomain 2: Distribution; Subdomain 3: Wire Field Operations; and Subdomain 4: Control Ops. Key questions posed in this discussion included: "Would the utility need to own new emerging assets inserted into the distribution system, beyond current assets?" and "Who would manage and maintain those assets? Would this be a new service and source of revenue for the utility?"

Domain 4: Metering

This domain features two subdomains: Subdomain 1: Meter & Associated Field Systems; and Subdomain 2: Meter Field Operations. "Communications and Data" should apply to both subdomains. Where the meter fits in the business process would become one of the key questions in Phase One. Emerging legal issues and business model constraints would need to be examined. Customer metering functions are seen as a part of the smart grid, but revenue metering should be distinguished from allowing remote control of appliances. Austin Energy went live with a new metering system in August 2009 and a new digital billing system in 2011, making time of use pricing (perhaps even real-time pricing) technically possible.

Domain 5: Distributed Energy Resources

The four subdomains include Subdomain 1: ES (PHEVs are designated as a subset of storage); Subdomain 2: DR; Subdomain 3: DG; and Subdomain 4: EE. A key distinction should be highlighted between DR and EE. EE is passive (nonintelligent and nonresponsive), while DR is interactive (it can be monitored and controlled via communications in the smart grid).

Domain 6: Customer

This domain is different from the five that precede it: All the other domains speak to physical systems and design, but this domain talks about users or customers. This domain would become quite crowded after much discussion, with specific definitions added for all who use the electricity produced and distributed by Austin Energy (beyond the traditional residential, commercial and industrial rate classes typical in utility systems)—Subdomain 1: Commercial; Subdomain 2: Residential; Subdomain 3: Industrial; Subdomain 4: Builders (green developers); Subdomain 5: Energy Providers (companies that will emerge to offer new retail energy services); and Subdomain 6: Government. Regarding Builders, it was noted that a new set of codes would be needed for builders and developers to incentivize and create new architectural concepts.

PSAF as Integration Tool

The PSAF thus served as the graphic representation of the domains and subdomains that constitute the supply chain of Austin Energy's market elements, infrastructure, and systems. The PSAF also became a tool to integrate the work of the other Pecan Street Project teams and served as a graphic representation of system integration. The PSAF diagrammed the systems and technologies and integration points that comprise the new paradigm, in both the current state and the future state (visionary), from operational, systems, and technology perspectives. Put another way, the architectural framework, with its components of domains, zones, stacks, and integration dimensions, outlines the systems that comprise the architecture and addresses the technologies that enable those systems. While domains are the major groupings within the framework and zones are the means to divide the utility's operations and functions, for example, the retail operations and wholesale operations. Stacks are the elements of interaction, including people (employees and customers), software, hardware, communications, security, and energy. Finally, integration dimensions are touchpoints between the domains.

Next, Team 7 began discussion using predetermined *initial questions* provided by the Strategy Implementation Team and the Governance Board, a process element that all teams followed. Team members were asked to provide individual answers to the questions, which the project manager collated and analyzed, creating the kernel for initial group discussion and providing a valuable team-building exercise. After two months into the process, Team 7 members had completed the questions and answers, augmenting the list as more questions arose in discussions, finally doubling the original set of questions.

Day-in-the-Life (DITL) Scenarios and Use Cases

The team took another two months to develop and discuss DITL scenarios, disregarding any previous notions of current technical capabilities and ultimately reorganizing the scenarios by subject area. The value of a DITL scenario lies as much in the development of the scenario as in its contents—it's as much about the journey as the destination. A DITL scenario provides a step-by-step evaluation of a particular change issue, revealing both positive and negative impacts that may have otherwise gone unnoticed until much later.

Next, the team used the best of the DITL scenarios to build nine use cases according to a standard template. Use cases differed from the DITL scenarios as more formal, higher-level evaluations of a potential change in operations or systems. The use cases were compared against the PSAF to ensure that each had the appropriate domains and subdomains in the right places with the right integration dimensions. This collaboration process was used to elicit new elements to the PSAF and help the team to rethink its concepts on current capabilities and what could be possible and when. Throughout the exercise, focus was maintained on desired functionality and convenience for the consumer.

Before getting under way with scenario discussions, however, the team spent some time on exercise design. First, the team saw patterns that would allow grouping the scenarios, for instance, the scenarios described adding devices and/or processes to the grid in three logical groupings. A *resource* primarily describes the supply side, in this case, new DG (e.g., solar PV system), so these scenarios described how new generation is treated in evaluating grid options. A *load* on the other hand, describes the demand side (e.g., smart appliance) that is incorporating far greater DR functionality through such technologies as HEMSs. Finally, the category *both* described storage (e.g., EV), which is the most disruptive of all, since it can be either a resource or a load depending on where it is in the charge or discharge cycle.

Another way to consider scenarios involves taking either *system* or *user* perspectives. Each scenario would require a description and notation of impacts on other Pecan Street Project teams. These criteria comprised a matrix/template for future discussion. Next, solution scenarios divided along two major alternatives: a *simple tech* solution would be low-tech, based on rules and processes that elicit the human feedback and behavior changes needed to accommodate a norm. In contrast, the *smart tech* approach would be high-tech, highly flexible and intelligent, characterized by tools and technologies that enable individual events. Scenarios were culled to produce use cases.

Other Smart Grid Planning Tools

The completion of use cases that correlated with the Q&A and the DITL scenarios marked the end of the first part of the team's work, whereby it was able to move on from analysis to integration and synthesis, as detailed in Figure 5.2. To determine the issues and impacts of each of these use cases on the energy and information systems within the local energy ecosystem, the team devised a content collection matrix, which mapped each use case onto a system architecture matrix. Near the end of the process, the team developed a list of business ideas and recommendations, which it then mapped onto an idea assessment matrix.

Along the way, the team gathered unaddressed items in a *parking lot list*, which it reviewed near the end of the process, assigning relevant items for further analysis. The team also met periodically with other teams to gather feedback and coordinated with the strategy implementation team along the way. The team submitted monthly *interim reports* to the Governance Board throughout the seven-month Pecan Street Project Phase One journey.

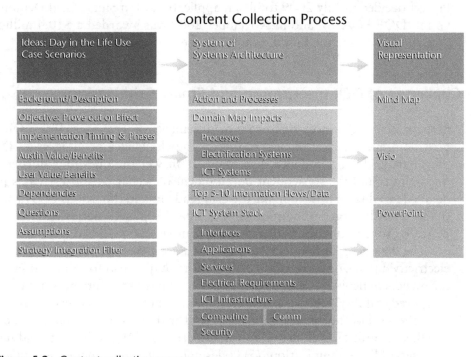

Figure 5.2   Content collection process.

Flexible Planning

Team 7 demonstrated a great degree of flexibility along the way to accommodate internal and external changes. For instance, as referenced earlier in this chapter, a great opportunity arose near the end of Team 7 discussions and deliberations when a new resource, the draft NIST Smart Grid Interoperability Standards Project, was brought on line as the fruition of earlier industry work. Back in 2007, NIST had been given primary responsibility by the federal government under the Energy Independence and Security Act (EISA) of 2007 to coordinate development of a framework, including protocols and model standards for information management to achieve interoperability of smart grid devices and systems, which was subsequently termed the NIST Smart Grid Interoperability Standards Project (see also Chapter 6). Since this report came out, NIST has used the report and the supporting organization to help guide development of smart grid interoperability standards. Team 7 determined that the NIST document would be a good companion to its team report and after an extensive review, mapped relevant use cases from its report against the NIST draft.

In another instance of rapid adaptation to changing circumstances, the Pecan Street Project planners had to adjust to the stimulus bill ARRA, which was passed only after the Pecan Street Project had launched and Phase One work had begun. Suddenly, with the passage of ARRA, a new opportunity for funding became available, so the team shifted from planning for potential bond issues to planning for potential smart grid grant opportunities. While Austin Energy filed an application for Smart Grid Investment Grant (DOE FOA 58) based on its work with the Pecan Street Project, it was not chosen as a grant recipient. Nevertheless, the Pecan Street Project decided in July 2009 to file an application for a Smart Grid Demonstration Grant (DOE FOA 36) and in November 2009 was awarded a $10.4 million grant that would become the foundation for Phase Two (see also Chapter 6).

The Challenge of DER Integration and Smart Grid Design

Perhaps one of the most significant achievements of Team 7 in Phase One was to provide a much richer understanding of the challenges of integrating DER into the smart grid. The in-depth discussion to answer the questions during the first few months and the subsequent development of scenarios and use cases provided detailed insights on how to make the grid ready to add thousands of EVs and incorporate a vastly larger amount of rooftop solar PV panels than is currently contemplated on today's distribution grids, for example.

The details of DER integration matter a great deal to the utility providing the electricity and ensuring reliability of the distribution grid for all the users, not just the owners of new DER assets. DER challenges must take into account the penetration, predictability, relative distribution, and finally, usage patterns that will determine the level of the integration challenge. The dimensions of challenges presented by DER integration include: (1) infrastructure cost; (2) grid impacts and reliability; (3) safety; (4) communication; (5) codes and regulations; including siting requirements; (6) variability; and (7) security.

The *economic* challenge for utilities will be to devise ways to avoid the inherent threats of DER integration, such as unplanned and irregular daytime charging, clustering, and revenue attrition as EVs and other DER become ever more common. Avoiding a blown transformer by adding distribution system capacity or upgrading a substation to accommodate such clustering of EVs will likely cost more than the utility could ever earn from the electricity alone that it will sell. Consider that utilities will earn relatively little from EVs that charge mostly at night, when electricity is priced by new TOU rates that may be as low as 2 to 3 cents/kWh. Special EV rates are likely to be required in the long run.

Current grid design matters immensely. In general, a single transformer in the United States may service only four to 10 homes, but in Canada and Europe, which have a different grid design, a larger transformer will serve as many as 20 to 30 homes, often even more.

Transformers and feeders will need to be upgraded and feeder management capabilities will need to be automated to accommodate EVs at any scale in order to minimize *grid impact*. Automating a distribution feeder to accommodate EVs will typically require adding three switches (at an approximate cost today of over $5,000 each). Beyond the distribution substation, protection gear such as relay protectors are not currently present to protect against adverse impacts from EVs.

The utility must work with the emerging EV manufacturers to enable scheduling capability for EVs, so the utility can control these events in real time, driving a need for IP networking infrastructure throughout the grid. Put another way, just as the early grid was designed to meet peak load on a macro basis, the emerging smart grid must be hardened to meet peak load at the edges, not only for safety, but for reliability and economic life of the grid.

As part of the new grid design, for instance, DER integration strategy now must encompass the concept of *islanding*, which would enable the grid to disassociate below the feeder level under certain conditions. Fortunately, DG technologies provide for sustainable islanding. A single control center today does not accommodate multiple islanding scenarios as a regular daily occurrence, so enhanced smart grid control will become ever more necessary.

Harmonics measuring and monitoring and modulation will become much more important to grid managers when integrating DER, not only for grid balancing, but also to ensure long equipment life. *Communication* capabilities of the grid will need to be significantly enhanced. Adding a variety of DER devices with different load characteristics will create a richer, complex management challenge as power quality begins to fluctuate more widely. If the current range of voltage fluctuates between 116 and 124V for an average of 120V, the addition of new devices can be expected to expand that range by a matter of degrees, to say 108–135V. Inductive load in particular, whether from EV charging stations or solar PV inverters, generates reactive power at far greater levels. Furthermore, solar PV carries the potential to provide an additional burden for the grid, given the *variability* and unpredictability of its electricity production.

Until the grid has been redesigned and enhanced to accommodate significant amounts of DER, however, changes in *codes and regulations, including siting requirements*, are likely to be implemented as a practical way to ensure grid stability and harmony, by prescribing where, when, and how DER elements can interconnect with the grid.

Finally, the *security* aspects of DER must be recognized from the outset. As described earlier in this chapter, security is a key feature of the emerging smart grid, and integrating thousands of new DER elements to the grid carries a significant security risk. Security must be considered at each stage of DER integration.

In summary, DER integration will emerge as a far more complex issue as technology advances provide ever cheaper and more functional solutions to add to the grid. Team 7 discussions during Phase One revealed that the smart grid will be the solution to DER integration but that the rules of integration are only now being mapped out.

Phase Two: Demonstrating an Energy Internet

Phase One could be said to have officially concluded in March 2010, when the Pecan Street Project issued a set of 39 recommendations in a report now available on the Web site [6]. The purpose of Phase Two shifted from brainstorming to generating new ideas and recommendations to applied research to quantify grid impacts and evaluate solutions in the field.

A key issue for Pecan Street, Inc., the nonprofit formed at the outset of Phase Two, would be to define how it would add unique value without duplication of effort with other Austin organizations. Potential roles for the nonprofit included: (1) outside analytical assistance, including measurement and verification; (2) public relations for a sustainable energy future; (3) project management for projects that receive ARRA stimulus funding; (4) a platform for fundraising on behalf of universities and university researchers in areas where funding gaps exist; and (5) reaching stakeholder consensus on priorities.

The ARRA grant application for Pecan Street had come together in a matter of weeks in July, with a plan to demonstrate an energy Internet at a new urban-style neighborhood that had already been partly constructed in a recovered section of downtown Austin, on the site where the former Mueller Municipal Airport had once stood. The new neighborhood—and the Pecan Street Energy Internet plan—took the same name. The Mueller neighborhood would become home then to the Mueller Energy Internet project—a microgrid in the center of Austin that would serve as a living laboratory to test new energy concepts.

As the neighborhood project got under way, homeowners began to move into homes that had either been built according to LEED designs or had followed Austin Energy's Green Building guidelines. Either way, living efficiently with the help of clean technology has been a driving force from the beginning for this neighborhood. The residents could be described as pioneers in a new way of living and were likely to be more open to new approaches than the general population would be. New urban infill neighborhoods like Mueller offer great potential for smart grid demonstrations.

As planned, the Pecan Street demo at Mueller provided an opportunity to put many of the ideas and concepts developed during Phase One to the test. A demonstration project goes beyond brainstorm sessions, modeling, and opinion surveys to assess potential solutions to both complex technical issues and novel social arrangements. The project would ultimately grapple with multiple challenges of the emerging energy Internet, including the following:

- *Two-way electricity meters* that provide customers with real-time information, including mobile phone access, collect and transmit a range of data, including data from smart appliances located inside the home or office; communicate with in-home/in-store displays and/or local energy networks; and measure energy flowing from buildings onto the grid (e.g., from solar PV panels); then, they communicate the data both to the grid operator and to local energy networks.

- *Local energy networks* (home area networks or office area networks) with supporting software that provide customers control of energy usage down to the appliance level—even remotely over wireless devices—respond to DR protocols from the grid operator; manage household smart appliances' energy usage based on electric budget, environmental preferences, variable pricing information, and other metrics set by the customer; manage household water usage based on electricity budget, environmental preferences, variable pricing information, and other metrics set by the customer; integrate variable pricing information; and interact with variable inverters.

- An *environmental dashboard* for larger communities (e.g., neighborhoods and all residences and the microgrid) that integrates the information collected through the local energy network, informing customers about the environmental impact of their energy and water usage.

- *Smart appliances* that integrate with the local energy network, possibly including HVAC and other major appliances.

- *Variable inverters* for homes and offices with solar PV that can be adjusted by the grid operator.

- *Plug-in vehicle charging* and energy management systems that integrate into local energy networks.

- *Utility-level functionality* that manages DR through a microgrid energy Internet; accommodates, accounts for, and manages two-way energy flow; provides information and instructions to local energy networks; and integrates EV charging with solar and ES in the Mueller Town Center garage.

- *Open-source design* that promotes replication—the intellectual property technologies and systems developed by Pecan Street, Inc. will be open-source and freely available, creating protocols for innovators to test their technologies on the energy Internet and to introduce new products and services onto the energy Internet that meet system requirements and achieve system values (e.g., carbon-free clean energy and reduced water usage).

- *Integration of water/reclaimed water systems* with the deployed energy Internet.

- *DG,* including solar PV (panel or thin film, or both) and possibly solar water heaters, and storage, including thermal storage, battery technologies (e.g., lithium ion, lithium iron magnesium phosphate, metal air, and lead acid), and possibly ultra capacitor and fuel-cell systems.

- *Business model testing* to measure functionality with customers, private sector job creation, and utility finances.

Like many of the DOE smart grid demonstration grants, the Pecan Street Project at Mueller would focus on smart grid technologies and DER integration. However, this demonstration was also unique in many ways: (1) a commitment to use open-source standards, (2) integration of a water system with a smart grid system, (3) the inclusion of green building technology and the integration of changes to building codes, (4) the commitment to utility financial viability, (5) the inclusion of a high level of affordable housing (25%), and (6) the integration of native landscapes.

Pecan Street Project Lessons Learned

When Phase One of the Pecan Street Project launched in 2008, the national economy was not yet in a tailspin, the federal stimulus bill was not even imagined, and the field of smart grid projects remained quite narrow. The smart grid industry was poised to burst onto the stage, however, as became apparent with the release of two U.S. DOE ARRA grant programs in April 2009. By October 2009, the Pecan Street Project Phase One technical report was in the first draft stage, and the DOE Smart Grid Investment Grant (FOA 58) awards had been made, earmarking $3.4 billion for 100 Smart Grid Investment Grant projects. One month later, the Smart Grid Demonstration Grant (DOE FOA 36) awards were announced, identifying regional Smart Grid Demonstrations (including the Pecan Street Energy Internet) that would share a portion of $600 million in funding. Finally, EPRI, the research organization associated with the electric utility industry worldwide, announced a program in 2009 for as many as eight smart grid demonstration projects. In November of 2009 the DOE awarded $10.4 million to the project as a smart grid pilot. By January of 2010, Pecan Street Project, Inc., was formed and hired Brewster McCracken as executive director and it, received a $297,000 economic development grant from the Department of Commerce. The Pecan Street Executive Report was published in March of 2010. The new Pecan Street, Inc., emerged in the summer of 2011 with a new focus around a new smart grid lab and the ability to work with device manufactures and utilities across the nation to share its learning and newfound expertise.

Where do the Pecan Street Project and Austin Energy now fit in the context of the evolving smart grid discussion? In short, these two organizations have blazed a long trail, but execution and fortitude will determine if they stay there given recent events. Having achieved more than most electric utilities over the past decade in reinventing itself through incremental changes, progressive Austin Energy is now undergoing a dramatic paradigm shift. A recent management change and new rate case approved in 2012 has slowed down many smart energy projects and efforts and shifted focus to some of the old ways. With the prospect of moving away from its traditional role of distributing commodity kilowatt-hours one-way out past analog revenue meters to dumb appliances and relatively passive consumers, Austin Energy has before it an opportunity to start providing dynamic energy services and prices over a two-way smart grid that includes smart meters, smart appliances, and more active consumers. We only hope that the execution of the successful and proven smart energy plans continue.

Team 7 Recommendations

Team 7 brainstormed the following list of recommendations in the final days of Phase One.

- *Distribution system operator (DSO).* Over the next four years, as Austin Energy introduces a variety of distributed resources onto its grid, its responsibilities and operational functionality will move closer to that of ERCOT, its host independent system operator. Austin Energy may choose to become a DSO (i.e., Austin Energy will need to have systems and processes to manage market functions and the flow of energy across the grid, like those at ERCOT that manage the transmission grid).

- *Independent DG dispatch.* With the ERCOT region in the process of transitioning to a nodal market, which will account for energy transfers based on specific dispatch from node to node, rules will be needed within the Austin Energy distribution grid to define parameters for third-party DG dispatch on the Austin Energy distribution network, an activity potentially brought under the purview of the new nodal market. This independent dispatch scenario will need to be defined with much greater detail.

- *DR and DG zonal development.* DR and DG decisions help determine the strategic vision of the utility. Subdividing the service territory into DR zones and DG zones would help to ensure optimal distribution so that these resources support the utility's vision and operational requirements. The criteria for zonal siting would need to consider distribution congestion, economic development, and disaster recovery. Managing energy costs in schools, a key public policy issue, would argue to put those facilities at the front of the line to receive subsidies for DG and DR, and the utility would also benefit from owning DG on these facilities to facilitate disaster recovery.

- *Rates versus information.* Rates based on TOU, RTP, and critical peak pricing (CPP) help a utility differentiate its commodity kilowatt-hours by price in order to motivate consumers to shift their consumption to off-peak hours. However, if a new paradigm of providing energy services were to be adopted, alternative rates may be more appropriate. Real-time information, for instance, would educate consumers about their consumption and its impacts in order to change their consumption behavior, perhaps avoiding altogether the need for rate-based price signals.

- *Cannibalization and transition.* The realignment of a city-owned utility to embrace DG must be managed to ensure that such an alternative does not cannibalize the utility's revenues and city services that provide key revenue for the city budget. A utility choice to shift capital investment from central generation and distribution facilities to utility-owned DER could delay or avoid altogether private-sector DER investment and utility revenue dilution.

- *Decoupling.* Decoupling breaks the connection between energy sold and income earned by providing a return on existing capital investment and disincentives for ancillary expenses that may become obsolete in a DER environment. A new rate structure based on decoupling would feature a required component embedded in every customer's bill that covered fixed capital investment and an optional array of charges that associate specific costs with specific services.

- *Incentives versus mandates.* Mandates carry with them an element of coercion that is out of alignment with a more inclusive approach to energy provisioning, suggesting a shift to incentives. The utility that shifts to providing energy services should ask for new customer behaviors to provide a more efficient and cost-effective energy ecosystem.

- *Customer segmentation and differentiation.* A shift to TOU rates will impact different customer classes differently. For instance, SMB customers have few to no options to shift consumption from peak business periods that coincide with peak demand periods. Multiple programs will be needed to accommodate different classes of customers, and feedback loops will be needed to track the performance of such programs.

- *Last-mile communication.* Current last-mile communication options are inadequate to provide a complete solution to distribution utilities to communicate sufficiently throughout the service territory. The data requirements to implement the solutions envisioned in this report overwhelm the existing communication options, suggesting a shift from narrowband and stronger emphasis on energy-dedicated last-mile IP networks.

- *Standards and interoperability.* The adoption of standards will be required for the full vision of an energy ecosystem to be fulfilled. Interoperability is a requirement for a fully evolved, integrated, communicative advanced smart grid. The utility system will need to integrate all the islands that technology providers create with their proprietary technologies. NIST and other groups will need to continue to drive this overarching goal of full interoperability. The utility will need to become a system integrator, providing the API in the cloud to let distributed solutions work together in a functional ecosystem.

- *Change management and the economy.* Organizational and infrastructure change is facilitated, even inextricably linked by the presence of economic growth, which provides a utility an opportunity for incremental implementation. Systemic transition can focus on gradual replacement of old equipment and processes with new ones to accommodate economic growth according to a new set of standards.

- *ES as an asset.* Each type of ES—central, substation, community, and premises—has potential to open new opportunities for distribution operations. However, the regulatory treatment of storage must be resolved for this emerging resource to be effectively implemented and deployed in a distribution utility service territory. Resolution of the treatment of this emerging energy resource in the capital markets is needed for it to become a reliable resource for utilities. ES is limited to pilot scenarios while utilities wait for costs to come down, benefits to be proven, and capital risk scenarios to be resolved.

Conclusions on the Next Generation Utility

Moving from its traditional role (selling commodity kilowatt-hours over a one-way distribution grid out through meters and appliances to relatively passive consumers) to a new role (providing energy services over a two-way advanced smart grid and smart devices to relatively more active consumers) constitutes dramatic change and demands new approaches, new attitudes, and new terminology.

Austin Energy has long held a vision to become a next generation utility. The Pecan Street Project went a long way to defining in greater detail what a next generation utility will look like. On the other side of an SG2 transformation will be a utility not only far more reliant on an integrated fleet of diverse DER systems and a base of efficient, smart consumers, but also no longer subject to the risks of rising fossil fuel costs and rising costs from new carbon externalities. Moreover, *integration* will define the next generation utility—integration with its own system (SG2); with emerging DA, DR, and DER technologies; with its own consumer community; with the transportation infrastructure; and with the water infrastructure.

In hindsight, the documentation in Pecan Street Project Phase One of potential reform ideas and new insights and processes to help facilitate change helps to move our national discussion forward. Phase Two promises to reveal even more valuable insights as do the other ARRA Demonstration Grant projects, which should be completed by the end of 2015. The devil will be in the details, as they say, and each electric utility will have unique issues to resolve. Austin Energy served itself well with the Pecan Street Project.

In a poll released in September 2009 and again in 2010 [7], *Intelligent Utility* magazine and *IDC Energy Insights* ranked Austin Energy at "near genius" level, as the second-smartest utility in the nation, behind only Sempra Energy of San Diego. In 2011, Austin Energy dropped to third place in the *IDC Energy Insights* rankings. Austin Energy has had the motivation, the means, and the methodology to remain a leading progressive utility. The Pecan Street Project and the Pecan Street, Inc., pilot that followed offer Austin Energy and other utilities a model, as well as a treasure trove of lessons learned to show the way to an SG2 project.

Chapter 6 will provide a snapshot review of the current state of smart grid activity in the United States. We'll explore the advances in research and development and show examples of projects that are progressing along the road just described in this chapter.

Endnotes

[1]   Thomas Edison invented the incandescent light bulb or, more succinctly, discovered a filament that would last for a very long time. [Edison also preferred a direct current (DC) system of power plants located close to the load, so we may need to start recognizing him as the "father of DC" at some point.] George Westinghouse favored alternating current (AC) and recruited Nikola Tesla away from Edison's Menlo Park laboratory to co-invent AC generation, transmission, and distribution system design with step-up and step-down transformers that we still have today. Tesla also invented the electric motor some years later to expand the purview of early utilities from electric light companies to the "power and light" title we are so familiar with now. Samuel Insull, Thomas Edison's assistant, went on to raise tremendous amounts of capital and create a business model to finance the construction of

power plants and electric grids: holding companies or "trusts" owning regional investor-owned utilities. His system of interlocking trusts drove an economic boom in the 1910s and 1920s, but then Insull became public enemy number one when FDR was elected. FDR's attorney general made Insull the first example of his "trust busting," leading to passage of the 1935 Public Utility Holding Company Act (PUHCA).

[2]   http://cognexus.org/id42.htm.

[3]   The Pecan Street Architecture Framework reflects the lessons learned in the Smart Grid Architecture design process described in Chapter 4.

[4]   This EV use case is the inspiration for the similar use cases outlined first in Chapter 3 and later expanded from a different perspective in Chapter 7. The disruptive potential of EVs led us to look at this subject from different perspectives.

[5]   As interpreted in the Pecan Street Project, the term DER included both the relatively low-technology EE resource class and the higher-technology of DR. EE and DR were included because of their potential to lower total energy demand requirements and peak demand requirements with a focus on the built infrastructure, which consumes 70% of the nation's electricity, and energy consumption behavior. Buildings built before 1970 are notoriously inefficient, lacking in basic insulation and other EE fundamentals. Elsewhere in this book, DER includes DG, EV, and ES, but not EE or DR.

[6]   http://www.pecanstreetproject.org.

[7]   Results from the 2010 poll were published in the January/February edition of *Intelligent Utility* magazine http://www.intelligentutility.com/magazine/article/203209/ austin-energy.

Today's Smart Grid

In Chapter 5, we described how the work at Austin Energy expanded into a community-wide discussion on the prospects for an energy Internet, leading to the creation of the nonprofit organization, now called Pecan Street, Inc., as an ongoing research and development project in the Mueller neighborhood in Austin. Pioneer efforts throughout the industry, notably the extensive work at the Pecan Street Project in 2009, prominent in the first edition of this book, begin to recede in the rearview mirror as we write the second edition four years later. Still, the detailed discussion of future issues in Chapters 4 and 5 anticipates the broader smart grid discussion at the national and international level that emerged in 2010. Oriented first on technology adoption to optimize grid operations with the heavy influence of ARRA grant projects, smart grid in 2013 and 2014 began to acknowledge and contemplate the disruptive impacts of increasing market penetration of DER, the focus of the second half of this chapter.

The first part of this chapter tracks the development of the smart grid concept beyond a small coterie of individuals and industry insiders and the emergence of the term "smart grid" into the mainstream from 2010 to 2012, and onward to 2015 and beyond, when transformation entered our vocabulary, providing vital context to fully understand smart grid. Smart grid embodies a reinvention of the electricity grid, the fundamental infrastructure of the modern global economy, painting a picture on a broad tableau across the world. So it should come as no surprise that momentum and interest in the topic has exploded in the past few years. Notably, speakers and writers continue to "introduce" the term smart grid, giving it a multitude of meanings, but we respectfully refer the reader anew to the succinct definition in Chapter 1 and the discussion in this chapter. The future of smart grid, certainly, remains to be determined along the lines outlined in Chapter 5, developed in this chapter, and elaborated in Chapter 7.

Smart Grid Foundations: 2000–2012

The term smart grid, defined by Andres Carvallo on March 5, 2004 (see Chapter 1), was heard mostly in the relatively small circles of the utility cognoscenti; among experts at NREL and other DOE labs; at EPRI, GE, Cisco, and IBM; from industry pundits like Clean Edge; among early members of the GridWise Alliance (GWA); and perhaps among such early North American utility innovators as American

Electric Power, Austin Energy, CenterPoint Energy, Duke Energy, Oncor, PG&E, Salt River Project, Southern California Edison, SMUD, and San Diego Gas and Electric [1]. Of course, early smart grid activity was by no means limited to the United States, although that is the primary focus of this book. Notable early movers in smart grid circles internationally included DONG (Denmark), ENEL (Italy), North Delhi Power and Light (India), Country Energy (Australia), EnergyAustralia (Australia), and SPAN (Australia). The list of early adopters could go on and on; this attempt to name the early movers and shakers in smart grid brings to mind the old adage: "Success has many fathers, failure, only one."

Smart grid remained a visionary concept of technology implementation and integration for most utilities until a few years ago, its potential tied to advances in technology that would improve the grid. Earlier this decade, the general trend lines foretold the advent of the new smart grid era, but few could describe in detail the paths the story might take, and that remains true today. With a few exceptions, notably journeys of discovery such as the on-the-job training inside Austin Energy documented in Chapter 4, smart grid activity was limited to research, pilots, and planning until tech advances and government support (e.g., ARRA grants) drove wider adoption.

Looking back, we can see the signs (hindsight is indeed 20/20). Clean Edge discussed the potential of grid optimization in its *Clean Energy Trends 2003* [2], describing how EPRI, U.S. DOE, Cisco, and a handful of utilities including Exelon, ConEd, and Salt River Project had formed the Consortium for Electric Infrastructure to Support a Digital Society (CEIDS) [3]. According to its mission statement, CEIDS was an attempt [4]: "to develop the science and technology that will fundamentally transform the infrastructure to cost effectively provide secure, high-quality, reliable electricity products and services." The vision of EPRI and the founding organizations behind CEIDS has stood the test of time. However, while prescient, their 2003 vision [5]—"A new electric delivery infrastructure that integrates advances in communications, computing, and electronics to meet the energy needs of the digital society"—illustrates how the industry 10 plus years ago was entirely focused on utility infrastructure modernization, ignoring the looming transformation beyond the meter in buildings, homes, and EVs.

Among other noteworthy early milestones, we must mention the creation of the GWA in 2003 and IBM's early work in the development of the smart grid maturity model, which since 2009 has resided at Carnegie Mellon's Software Engineering Institute. Prior to 2009 in the United States, there was much more talk than action; efforts to build the smart grid resided principally in California, Colorado, and Texas. Investor-owned utilities and city-owned SMUD in California began the early days of AMI-led smart grids. Boulder, Colorado, was the site of the pioneer Smart Grid Smart City project by Xcel Energy. Similarly, Austin Energy explored the future with the Pecan Street Project and finished the first fully deployed smart grid in the United States in 2009. In Texas, Austin Energy was closely followed by Oncor, Centerpoint, American Electric Power Texas, Bluebonnet Cooperative, and City Public Service in San Antonio.

The smart grid starting gun [6] truly went off with the enactment of ARRA in February 2009 [7] with the federal government allocating nearly $4 billion in funding to support matching grants in two categories—investments and demonstrations. As grant awards were announced in late 2009, the emerging smart grid

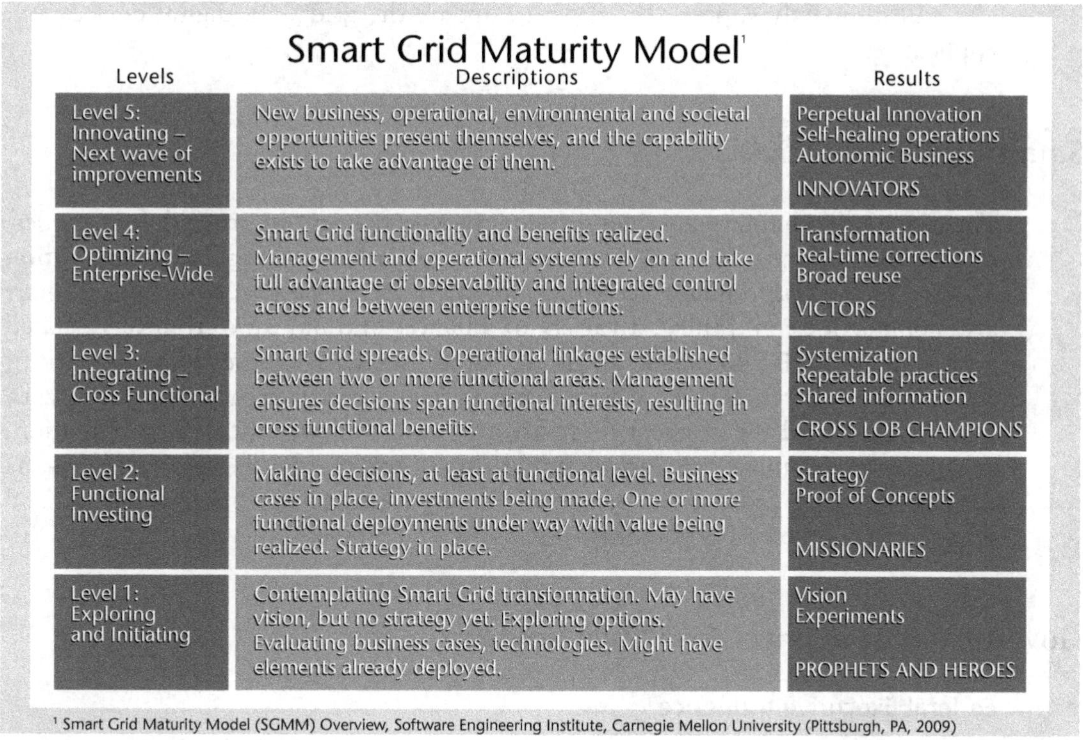

Figure 6.1   Smart grid maturity model.

landscape unfolded with winners receiving a kick-start to their projects and those not winning awards left to ponder next steps.

By 2010 the realities, costs, and complexities of building a smart grid were starting to become apparent to more and more stakeholders—in some cases, the bloom had begun to fall from the smart grid rose. Consumer groups challenged the smart grid value proposition of massive smart meter rollouts, first in California (PG&E), but later in more and more jurisdictions (e.g., Texas, Illinois, Maryland, and Maine). By the end of 2010, AMI was no longer the automatic first step to building a smart grid. Operational issues elevated grid optimization and DA applications as a viable alternative to AMI, providing a double advantage, first avoiding direct consumer backlash and second providing a more attractive business case to a more skeptical audience of regulators and consumers.

By 2011, the concept of a smart grid had become far more widespread, showing up in mainstream publications like *Scientific American* [8], *Newsweek* [9], and the *New York Times* [10]. At that time, although the smart grid was the most significant change to hit the fundamental electric grid infrastructure since its creation over 100 years ago, the actual implementation of smart grid projects remained limited to foundational smart grid projects. The discussion kicked into high gear, however, with standards, interoperability, and security front and center. NIST [11] stands out as the leader in this area, with its SGIP [12] and security documentation [13] developed in coordination with the GridWise Architecture Council (GWAC) [14]. In terms of interoperability, several more standards bodies also receive our attention in the following discussion, given the fact that a plethora of technologies will ultimately support the drive to update the grid with digital devices and applications.

Smart Grid Perspectives

The rest of this chapter examines different perspectives on smart grid: first, leadership at the national and state level through legislative and regulatory institution activity on smart grid issues; second, foundational U.S. standards to manage smart grid complexities; and third, a variety of alliances and industry groups associated with smart grid development, including helpful media resources to keep track of smart grid progress. Finally, a new section reviews recent trends associated with DER that expand the concept of smart grid to include business transformation, consumer engagement, organizational changes, and regulatory reform efforts, as the industry shifts from its early, singular focus on technology implementation and integration focused on grid modernization.

Government as a Smart Grid Stakeholder

Federal Executive Influence

Office of Science and Technology Policy (OSTP), Executive Office of the President

In November 2010, OSTP [15] issued a report titled Accelerating the Pace of Change in Energy Technologies through an Integrated Federal Energy Policy [16], providing a road map for federal government transformation of the U.S. energy system over the next two decades.

Federal and State Legislative Influence

Title XIII, EISA 2007

A discussion of federal legislative influence starts with a review of 2007's EISA [17] specifically, the smart grid section, Title XIII, which directs the task force to: (1) produce regular reports on the status of smart grid deployments nationwide (Section 1302); (2) carry out a program to research, develop, and demonstrate smart grid technologies and establish a smart grid regional demonstration initiative focused on advanced technologies (Section 1304); (3) establish a federal matching funds program (Section 1306); (4) submit to Congress a study assessing the effect of private wire laws on the development of CHP facilities (Section 1308); and (5) submit to Congress a study on the security aspects of smart grid systems (Section 1309). Three other sections: (1) outline federal policy toward smart grid (Section 1301); (2) direct NIST to set up a smart grid interoperability framework (Section 1305); and (3) provide direction to state regulators on appropriate oversight of utility investments in smart grid (Section 1307). Title XIII got the ball rolling on research, funding, and regulatory direction, but it was ARRA two years later that really launched major smart grid activities.

DOE National Laboratories

At least six of the DOE's 21 national laboratories have a special focus on smart grid (Figure 6.2) [18]. The National Energy Technology Laboratory (NETL) [19] in Morgantown, West Virginia, is working on upgrading the national transmission and distribution system by developing a nationally coordinated grid modernization framework. NREL [20] in Golden, Colorado, is the primary R&D lab for renewable energy and EE. Los Alamos National Laboratory (LANL) [21] in Los Alamos, New Mexico, focuses its smart grid activity on the ITC aspects of grid design, grid control, and grid stability. Sandia National Laboratory [22] in Sandia, New Mexico, has projects in renewable ES and solar energy technology [e.g., Solar Energy Grid Integration System (SEGIS)]. Lawrence Berkley National Laboratory (LBNL)]23] next to the UC Berkley campus outside San Francisco is researching EV technology and has developed an open source communication specification that supports automated DR (OpenADR). Finally, Pacific Northwest National Laboratory (PNNL) [24] in Richland, Washington, brings a special focus to the environmental side of smart grid.

SGIC and SmartGrid.gov

DOE set up a process for gathering and disseminating smart grid data, leveraging the online SGIC [25], DOE requires smart grid grant recipients to provide information from their projects to NREL, which funnels the information to the SGIC. The SGIC is also open to information contributed from other smart grid projects and sources.

DOE-Funded Smart Grid Projects

With the enactment of ARRA in February 2009, the DOE issued two funding opportunity announcements (FOAs): FOA 58 for large-scale investment grants and FOA 36 for smaller demonstration grants. One hundred FOA 58 matching investment grants totaling $3.4 billion in multiple categories were awarded in late 2009,

DOE National Labs and Smart Grid

- National Energy Technology Laboratory (NETL)
- National Renewable Energy Laboratory (NREL)
- Los Alamos National Laboratory (LANL)
- Sandia National Laboratories (Sandia)
- Lawrence Berkley National Laboratory (LBNL)
- Pacific Northwest National Laboratory (PNNL)

Figure 6.2   Smart grid DOE national labs.

with a focus on AMI and "integrated" projects (projects with multiple types of smart grid devices and applications). The grants followed the 80/20 rule: 25 large projects ($20–200 million) received 80% of funding and the remaining 75 small projects (less than $20 million) received 20%. Soon thereafter, funding for 10 FOA 36 demonstration and storage grants were awarded, totaling another $600 million. As these DOE-funded projects conclude, they have been tracked at SGIC. Given the major momentum provided by ARRA grant projects to the development of smart grid in the United States, many have pondered where future momentum will come from. What will fill the gap left by the receding ARRA? In this chapter, we propose that DER is providing the needed impetus to drive transformation, the next wave of smart grid.

Renewable Energy Standards (RES) and Renewable Portfolio Standards (RPS)

State legislative activity on smart grid focuses principally on renewable energy and EE. RES provide utilities a mid-range target, encouraging a shift to a more sustainable energy portfolio and economic development and job growth based on clean energy. To date, over 30 states and the District of Columbia have such standards [26]. RES and RPS are important drivers of the emerging DG market, specifically rooftop solar PV. However, standards-driven renewable innovation may give way to new market drivers as discussed in the policy section below (see NY PSC REV).

Federal and State Regulatory Influence

The FCC [27] regulates interstate and international communications by radio, television, wire, satellite, and cable. FERC regulates interstate electricity transmission and wholesale electric transactions in interstate commerce [28]. State regulatory bodies—such as public utility commissions and public service commissions—regulate intrastate electricity activity, specifically investor-owned local distribution companies and, in areas where retail electricity service is now competitive, monopoly transmission and distribution utilities.

The FCC published its *National Broadband Plan* report in June 2010, with Chapter 12 devoted to energy and the environment, underscoring the finding that broadband is essential: "to lead the world in 21st century energy innovation [29]," citing four key ways that broadband will enable energy innovation. First, broadband will unleash energy innovation in homes by making energy data readily accessible to consumers; consumers can make simple changes with feedback and smart appliances can automatically connect with the grid. Second, different broadband technologies will work together to make the grid more reliable and efficient, advancing innovations in renewable power, grid storage, and vehicle electrification. Third, broadband will improve the EE and environmental impact of the information and communication technology (ICT) sector, specifically data storage centers and server farms. Finally, broadband will make the transportation sector safer, cleaner, and more efficient with real-time traffic information systems and navigation tools and Web conferencing and telecommuting to substitute for mass transit commuting and transportation.

The need to educate state policymakers on smart grid options remains immense (the average tenure of a state commissioner is just over three years). Balancing competing priorities of maintaining and upgrading a reliable grid is an overwhelming challenge for regulators. FERC has actively promoted adding new types of energy resources to the grid [30]. Former FERC chairman Wellinghoff long stressed the value of DR to the wholesale market and the positive potential of EVs for the electric industry. National Association of Regulatory Utility Commissioners (NARUC) [31] represents the state commissions that regulate electricity, telecommunications, water, and transportation, making it well-positioned to lead smart grid. The NARUC-FERC Smart Grid Collaborative [32] helps policymakers find their way through the maze of options by providing a forum for state and federal regulators to share viewpoints on a variety of technical, political, and economic issues.

Industry Standards and Security

SGIP

Under EISA 2007, the U.S. federal government gave NIST primary responsibility to coordinate development of a smart grid framework with protocols and model standards for information management to achieve interoperability of smart grid devices and systems. NIST created the SGIP [33] in 2009 to engage smart grid stakeholders for technical assistance in assessing standards needs and developing the smart grid interoperability framework. In 2010, NIST issued its first release of a smart grid interoperability framework and roadmap for its further development. From 2010 to 2015, SGIP has steadily grown into a valuable institution for ensuring interoperability of technologies and systems, but also for providing a valuable knowledge base on such critical topics as cybersecurity, Green Button, safety standards, and emerging smart grid concepts.

Smart Grid Architecture Committee (SGAC)

The SGAC developed a Smart Grid Conceptual Model [34] as a tool for discussing the structure and operation of the power system. The conceptual model defines seven domains as well as actors, applications, associations, and interfaces that can be used in the process of defining smart grid information architectures, such as the combined conceptual reference diagram.

Domain Expert Working Groups (DEWGs) and Priority Action Plans (PAPs)

SGIP created five DEWGs to focus industry domain expertise: (1) transmission and distribution; (2) building to grid; (3) industry to grid; (4) home to grid; and (5) business and policy. PAP are tools used in SGIP to support the analysis and application of standards to the smart grid use cases. The 17 PAP Working Group Management Teams inside SGIP develop PAPs to address either a gap where a standard or standard extension is needed or an overlap where two complementary standards address some common information but are different for the same scope of an application.

Industry Standards Groups

NIST and SGIP bring a multitude of standards bodies together, but a variety of other standards groups are more focused on a single set of standards, described below.

U-SNAP Alliance

"U-SNAP" is an acronym for utility smart network access port. The U-SNAP Alliance [35] promotes a universal solution that enables any home area network standard to use any vendor's smart meter as a gateway into the home, without needing to add more hardware in the meter.

IEEE P2030 (Smart Grid Interoperability of Energy Technology and IT Operation)

This standards group provides guidelines for smart grid interoperability between the grid and end-use applications and loads. The IEEE P2030 [36] standard addresses interconnection and interfacing frameworks and strategies with design definitions needed for grid architectural designs and operation.

Open Smart Grid (OpenSG) Subcommittee

The OpenSG Technical Subcommittee [37] was created to foster enhanced functionality, reduce costs, and speed AMI and DR adoption through the development of an open standards-based information/data model, reference design, and interoperability guidelines.

Open ADR Alliance

The Open ADR Alliance [38] has become a critical element to introducing the demand side into grid operational equations. With OpenADR, device manufacturers are ensured interoperability, accelerating market adoption for their standards-based products. However, OpenADR serves another critical function, easing the integration of the traditional supply side operations with the emerging demand side complexity.

ZigBee Alliance

The ZigBee Alliance [39] mission is to enable reliable, cost-effective, low-power, wirelessly networked monitoring and control products based on an open global standard to provide the consumer with ultimate flexibility, mobility, and ease of use by building wireless intelligence and capabilities into everyday devices. This standards-based wireless platform is optimized for the unique needs of remote monitoring and control applications, including simplicity, reliability, low cost, and low power. This industry alliance will be instrumental in defining the communication protocols most likely adopted for short-range wireless communication among household devices in the emerging HEMS market, since many smart meters are adopting ZigBee compatibility (e.g., the deployed Smart Energy V1.0 specification with millions of units installed).

HomePlug Powerline Alliance

The mission of the HomePlug Powerline Alliance [40] is to enable and promote interoperable, standards-based home PLC networks and products, ranging from very high-speed technology capable of carrying multiple high-definition AV channels to low-speed, low-cost, low-power consumption PLC for home automation. Home-Plug is currently the leading technology candidate for in-home PLC.

Wi-Fi Alliance

As with ZigBee and HomePlug, the mission of the Wi-Fi Alliance [41] is to certify Wi-Fi products (such as radios based on IEEE 802.11.a, g, and n), promote Wi-Fi products and markets, and create industry standards and specifications. The Wi-Fi Alliance has created a new working group to more fully develop smart grid specifications and recommendations.

Consumer Interest Groups

On November 14, 2010, prior to the beginning of the NARUC Annual Meeting in Atlanta, Georgia, the Critical Consumer Issues Forum, entitled "Focusing on Smart Grid from the Consumer Perspective," jointly sponsored by NASUCA [42] (consumer advocates), NARUC (state regulators), and the Edison Electric Institute (electric utilities), hosted a discussion on the importance of consumers. A 2010 Accenture study, [43] "The New Energy World: The Consumer Perspective," showed that consumer adoption will be a key to the success of utility metering, conservation, and DR programs and that substantial budgets will be needed to induce the changes in consumer behavior on which utilities have been counting.

Smart Grid Consumer Collaborative

In March 2010, a group including Best Buy, Control4, Ember, GE, GWA, IBM, and NREL announced the formation of the SGCC [44] to promote the improved understanding of consumer needs in the smart grid universe. The SGCC has grown into a powerful advocate in the industry to promote and stimulate dialogue concerning consumer engagement in smart grid.

National Association of State Utility Consumer Advocates (NASUCA)

For more than 30 years, NASUCA [45] has provided a forum for organizations that represent utility consumers in regulatory and court proceedings. At the time of this writing, membership has grown to 44 consumer advocate organizations from 40 states and the District of Columbia. In 12 states, consumer advocacy is handled by state attorneys general, while in 29 other states, consumer advocacy offices have that role, with directors appointed by governors. Traditionally, NASUCA member organizations have challenged rate increases in adversarial rate cases.

Electric Industry Interest Groups

The U.S. electric utility industry is fragmented, composed of regional utilities formed to address regional needs in this vast, widely diverse country. Domestic transmission and distribution grids (3,121) deliver power to meters (about 140 million) that represent U.S. citizens (over 300 million). Those 3,121 grids break down as follows: 219 investor-owned utilities, with larger service territories and most of the major population centers; 2,010 public power utilities, with service territories mostly coterminous with city boundaries; 883 cooperative utilities, which are member-owned and were the last to the game, filling in territories that were not served by IOUs and MOUs; and nine federal power agencies. In competitive markets, there are additional retail marketers—in Texas alone, more than 150 retail electric providers (REPs) sell to commercial and residential customers over the existing power distribution networks owned and managed by Oncor, Centerpoint, and American Electric Power Texas. The associations in this section represent the industry and these different segments.

EPRI

EPRI [46] has taken an active role in fostering a collaborative environment among the nation's utilities and other interested parties to support smart grid research projects and large-scale demonstrations to ready supporting technologies for commercial operation as the smart grid develops. The EPRI Smart Grid Demonstration Initiative is a five-year collaborative research effort focused on the design, implementation, and assessment of field demonstrations to address prevalent challenges related to integrating DER in grid and market operations to create a VPP.

EPRI Inverter Program

In 2009, the EPRI Photovoltaic & Storage Integration Program (P174) [47] began to study ways to help manage a high DER penetration, with one research area specifically focused on the communication aspects of DER. By mid 2009, this research had led to the launch of a broad industry collaborative to identify a common means for smart, communicating inverters to be integrated into utility systems. The DOE, Sandia National Labs, and the Solar Electric Power Association (SEPA) joined with EPRI to help steer the research project.

GWA

As described at the beginning of this chapter, the GWA [48] coordinates smart grid organizations, principally by facilitating activities between stakeholder groups and by developing a sound foundation of educational and policy materials and acting as the go-to industry representative for government policymakers and the press when it comes to all issues associated with smart grid.

GWAC

Neither a design team nor a standards-making body, the GWAC [49] is a group of industry leaders, formed at the direction of the DOE, to help shape the architecture of the emerging smart grid, with a focus on interoperability and standardization .

Association for DR and Smart Grid (ADS)

The original DR Coordinating Committee (DRCC) and later, the DR Smart Grid Coalition (DRSGC) helped to raise awareness of DR and of the potential of DR as an emerging smart grid resource. Now renamed the Association for DR and Smart Grid, the ADS [50] hosts the annual national town meeting on DR and smart grid every year in Washington, D.C. [51]. In March 2010, Pike Research [52] predicted rapid growth for the DR industry starting in 2013, going as high as $8.2 billion by 2020. In June 2010, FERC staff published the National Action Plan on DR [84]. Today, DR is recognized as a key element of DER and has tremendous potential to transform the grid, utility operations, and consumer engagement.

Solar Energy Industry Association (SEIA)

The Solar Energy Industry Association (SEIA) [53] represents the solar industry.

EVs

Early EV efforts were driven by such industry groups as Plug In America [54], the Electrification Coalition [55], and the Intelligent Transportation Society of America (ITS America) [56], which promoted the development of the budding EV industry. As the industry matured, EV manufacturers and charging station companies took up the leadership mantle as described below.

ES Association (ESA)

The ESA [57], an international trade association promoting the development and commercialization of ES technologies, has been hard at its task for nearly two decades.

Edison Electric Institute (EEI), American Public Power Association (APPA), and National Rural Electric Cooperative Association (NRECA)

As the association representing investor-owned utilities, EEI [58] provides smart grid online resources, reports, workshops, and focus at its conferences, roundtables and seminars. At the EEI Roundtable in October 2010, EEI presented a Smart Grid Scenario Project Update [59], showcasing results of its two workshops in 2010 on smart grid, held in Washington, D.C., and Los Angeles. The key takeaways from the workshops on potential scenarios included the following: (1) that technology will have a transformative impact on the utility industry; (2) that new market entrants will be strategically positioned between customers and their utilities, leading to customer disintermediation (attractive energy packages from vendors will bypass and displace the utility, leading customers to become less reliant on utilities); (3) that a new customer culture (think iPhone) will exist, where real-time information becomes increasingly available via multiple technology platforms; (4) that the disruptive impact of smart technologies will occur to incumbent utilities as with telecom companies, regardless of any attempts at regulatory protection; (5) that a great many more customers will become less dependent upon utilities; (6) that supply resources will come increasingly from both traditional central station systems and distributed resources; and (7) that there will be multiple utility business models with varying probabilities of success depending on regional market structure.

EEI counterparts for other utility market segments include APPA [60], which represents the U.S. city-owned electric utilities, most of which are small distribution-only operations, and the NRECA, [61] which represents the U.S. member-owned electric cooperatives.

UTC

Having represented utility telecommunications issues and utilities since 1948, the UTC [62] has the perspective needed to contribute to the debate on a smart grid transition. From its Smart Grid Policy Summit in Washington, D.C., in April 2010, to its online UTC Insights, the smart grid focus of UTC blends technology and policy expertise.

National Rural Telecommunications Council (NRTC)

Supporting both electricity and telecom rural cooperatives, the NRTC [63] provides technology and procurement support to facilitate telecom solutions.

Environmental Interest Groups

Environmental Defense Fund (EDF)

Formed in 1967, the EDF [64] has long been an outspoken advocate for the environment. Devoting significant resources and leadership to the Pecan Street Project in 2009, EDF developed a strategic interest in electric utility smart grid projects, promoting clean energy best practices to the utility sector aligned with its own environmental goals. This new EDF utility division has gone on to actively support grid modernization throughout the United States.

Smart Grid Media and Events

Smart Grid Publications

A variety of publications appeared in 2007 and 2008; most provide daily updates via e-mail. A leading online newsletter available at no cost is Smart Grid News

[65], a pioneer in smart grid reporting and analysis. Energy Central publications include IntelligentUtility [66 an online daily that has grown into a must read publication, and two others, EnergyBiz [67] and RenewableBiz [68], which bring more focus from the business side of the equation. Other publications come at smart grid from specific perspectives. For the utility viewpoint, Public Utilities Fortnightly [69] consistently delivers valuable information as do Electric Light and Power [70] and Power Grid International [71], perhaps with a stronger engineering focus, as does the bimonthly IEE Power and Energy Magazine [72]. Insights on the renewable energy business can be found on the daily online e-mail newsletters Renew Grid [73] and Renewable Energy World [74]. Early in the second decade of the new millennium, four other publications emerged to become daily must reads. First, Fierce Energy [75] and Fierce Smart Grid [76] (which voted Andres Carvallo and John Cooper to its list of "most influential in energy" in 2011 after the publication of The Advanced Smart Grid, First Edition). GreenTechMedia [77] is a vital source of daily information on the ever busier sectors of DER, which it has dubbed The Edge. (It's catching!) GreenBiz.com [78] helps to tie all the emerging news together. Finally, Utility Dive [79] has become a good online resource to track industry trends. The authors gratefully acknowledge the tremendous service offered by all these media, but especially Smart Grid News and these last four groups, all vital resources we leaned on in compiling the updates in this chapter. To be more complete, this short list of publications must be complemented by publications from the myriad industry associations detailed in this chapter.

Smart Grid Events

Smart grid events in the United States and abroad continue, but event exhaustion looms with these types of events; attendance wanes as continued discussion but less action than anticipated causes attendance and revenues to drop. Policy formation is indeed helped by events that seek to publicize the basic facts and bring people together to stimulate dialogue and policy formation. Among the events that have gained a significant foothold in the smart grid space are Grid Week [80]; the GWA meetings; Distributech [81]; and annual meetings of UTC, EEI, NARUC, NRECA, APPA, and ADS. A new event platform, the Energy Thought Summit (ETS), emerged in 2013 to capture the imagination of a much larger ecosystem that uses and cares about all things energy.

From Smart Grid to Transformation: 2013–2015

Until 2012, smart grid had remained principally focused on technology innovations and grid optimization, as ARRA grants proceeded apace. Far more attention was paid to smart meters than to solar PV. However, by 2013, and increasingly in 2014, two clear themes emerged, based on distinct grid and edge perspectives. Inside utilities, we saw more discussion of grid value and the rising risk of economic and operational disruption. Outside the utility industry, alternatives to grid-supplied power fueled a growing chorus of innovation and its potential to upend the traditional utility industry, creating vast new value potential for outside stakeholders. So a life-threatening existential challenge for an incumbent (i.e., utility death spiral) became an economic boon when viewed through the lens of nonutility players seeking to move into the electricity sector. At the dawn of 2015, advanced smart grid has expanded dramatically—in multiple directions. However, one trend is crystal-clear: Changes will be dramatic...the age of transformation has arrived.

While grid optimization activity has continued apace over the past three years, we've chosen to focus our attention from 2013 to 2015 on disruption and transformation. What do we mean by transformation? In short, rapid maturity in distributed energy technologies (i.e., DER) driven first by maturing technologies and scale production, then supporting policies, and more recently by new business and financing models, has caused prices to tumble and value to increase multifold. This unexpectedly rapid development demands a response from the incumbent utility industry, even as it opens the door to new players. In the following sections, we'll provide snapshots of technology trends in DER and then edge power policy trends, as we named this trend five years ago, showcasing DER developments and institutional responses over the past few years. Finally, we'll look at disruptive phenomena from the business model perspective, specifically DER business trending and growing indicators of new business models trending as utilities seek to accommodate the rise of DER and edge impacts.

Trend 1: Technology Innovation

AMI matures.   An October 2013 article captured the challenge of AMI companies as large-scale AMI deployment opportunities decline without ARRA grants in a maturing U.S. market. Beyond chasing new metering business abroad, companies like Itron, Landis+Gyr, and Silver Spring are diversifying their offerings and moving into higher margin service offerings and new areas, including DER development [82]. Echelon has abandoned the smart meter altogether, selling off its smart grid business to focus on the Internet of Things [83]. Playing in a highly dynamic but slow adapting utility market is a huge challenge for meter deployment companies that rode the early smart grid wave. Late in September 2014, Silver Spring announced layoffs that came despite efforts to innovate beyond the meter, including networked streetlights and smart cities [84]. Similarly, Itron announced layoffs as well.

Making solar PV dispatchable.   Distributed solar PV (DSPV), our term for smaller residential and commercial solar PV systems, has experienced dramatic adoption as panel costs have dropped. However, the current system model of inverters connected to a net meter limits DSPV system value, opening the door for improvement with a more technically advanced inverter. A smart inverter would offer the utility better control of DSPV systems and enable users new value options. This trend appears to be gaining steam, for example, in Arizona, an area with high solar insolation (amount of solar energy available) and great potential for PV, Salt River Project (SRP), the utility that serves Phoenix, plans to retire 750 MW of coal and then stimulate enough DSPV to replace the lost central generation. As detailed in an August 2014 online post [85], SRP must develop and adapt new inverter technology to help DSPV fit into the utility business model of controlling and dispatching generation resources to match load demands for power. Smart inverters have the potential to be a new smart connected network element on an advanced smart grid.

Solar plus storage: grid substitute. In March 2014, Solar City introduced the idea of combining Tesla batteries with its DSPV systems to enable emergency backup for residential systems and peak shaving for commercial systems [86]. What if it could do more? When these systems are networked to the cloud, with data processing on-site, they represent far more than disaggregated power plants. These empowered, networked systems represent tremendous potential, from replacing peaking power plants to providing ancillary services to the grid, from providing wholesale power to meeting market demands to providing an alternative to net metering. As storage costs decline, the full potential of this bundled approach should be more and more in the news.

EV trending up—watch California. We know, this storyline gets a little stale, but it is true...as California goes, the nation (most generally) follows. If that's the case, we should get ready to see EV penetration numbers rise, as California passed the milestone of 100,000 EVs on the road in 2014 (10,000 of those in San Diego alone, where EV drivers have access to over 800 public charging stations) [87]. California EVs represent 40% of the U.S. total. Why is that? According to the article, "Why Electric Cars Are Selling in California: They're Free," cars are flying off of showroom floors in California because the winning combination of tax credits, state rebates, employee subsidies, and free fuel make EV ownership almost irresistible.

EV trending up—not so fast. But we're not all California, as it were. According to an August 2014 piece on CNBC [88], experts are widely split on predicting the uptake of EVs. First, there's something called "sizzle and fizzle," where trends rise and fall—sustaining a trend is a remarkable challenge, especially if rebates are required for very long. Second, battery technology has to get better to overcome the trade-offs between range and battery weight. Finally, don't expect internal combustion engine (ICE) vehicles to go gently into the night. As ICE technology improves over time, the decision to switch to EV will be made more difficult by increased mileage and better economics from improved ICE technology.

Opening up to new ideas. It is important that the cost to develop new technology drops if it is to head off competition in time. And if the goal is to stimulate EV market growth by opening the door to the marketplace of new ideas, Musk did so in June 2014 when he made Tesla's patents on EV design publicly available [89]. The argument goes that lowering R&D costs for EV development will lead to a healthier EV industry, ideally heading off a budding fuel cell vehicle at the pass. With healthier EV producers, we may expect greater competition and a larger EV market, sooner.

EV and ES at scale. If EVs are to gain wider acceptance, another way is to attack the cost of one of the most expensive EV components—lithium ion batteries. Tesla raised eyebrows in 2014 with its widely publicized plans to build a huge facility to manufacture lithium ion batteries, with output directed at its Tesla EVs, but for other purposes as well. Five states in the Southwest actively sought to land this deal (Arizona, California, Nevada, New Mexico, and Texas). Finally by September, Nevada emerged as the winner, though some wondered if Nevada had given away too much to win this prize [90]. But what a prize! The factory will be huge, covering 10M square feet, an area equal to 174 football fields. It will open in 2017 and employ 6,500 workers in the suburbs of Reno. One article compared the dramatic economic impact in Nevada to that of the Hoover Dam nearly three generations ago. For our grandmothers and great grandfathers 80 years ago, a huge power plant construction project was remarkable. For us today, we may expect this huge factory to carry similar impact, certainly for regional revenue and employment, but also for the budding EV and ES markets. There's still much to this story to be told, but now that Nevada has been selected, it appears well on its way.

The EV/ES/PV connection.   The plot thickens when we string all these stories together. Cheaper batteries will not only help lower the cost of EVs. Cheaper, stronger lithium ion batteries will stimulate PV system sales when they are bundled together. And when we add in on-site computing capacity and communications, we begin to have a system with significant potential. As the numbers grow, the potential for aggregation becomes compelling. When that happens, the EV/ES/PV connection may prove to be game-changing [91].

Natural gas fuel cells.   In some quarters, natural gas is increasingly viable based on fracking and other technological advances. While a variety of factors may support or challenge such a statement, there is a class of new technologies standing ready to take advantage of available and affordable natural gas. New dispatchable devices like fuel cells and microturbines represent natural complements to a more diverse energy ecosystem challenged by intermittent solar PV and wind. When managed together in a complementary fashion, these technologies have tremendous potential to achieve balance locally, decreasing or eliminating the need for peaking power plants to balance the grid. A review shows steady progress in both technologies, but there is no billion-dollar microturbine company to match what Bloom Energy does to put fuel cells in the conversation. Rather, the technology proceeds with evolutionary improvements. Microturbines equipped with heat exchangers take advantage of waste heat to provide CHP with special applications at hotels, hospitals, and nursing homes, locations that share a significant need for domestic hot water. However, fuel cells, Bloom's in particular, are so far stealing the limelight from microturbines, as described below.

Fuel cell landmark sale.   In October 2013, Forbes reported that eBay's new data center in Utah would depend on fuel cells from Bloom Energy running on natural gas—the system has 30 Bloom Energy servers in all, fuel cells that provide 6 MW of generating capacity [92]. That's 75 percent of its 8-MW total load requirement. For now, eBay will get the rest of its power from the grid, but it has plans to get electricity from waste heat at a natural gas facility miles away. In short, eBay has found a way to clean up its act with a smaller carbon footprint that is at the same time more reliable. When this happens, the utility finds itself relegated to backup energy supplier.

Creative financing drives fuel cell sales. As with solar and wind purchase power agreements (PPAs), which are similar to a lease, Bloom is using creative financing through its Bloom Electrons program to remove buyer objections and stimulate sales. Notable for having raised over a billion dollars in corporate financing, Bloom announced another type of financing in July 2014 [93]. Bloom has lined up Exelon, which is committed to finance up to 21 MW of Bloom fuel cell projects, working closely with subsidiary Constellation's distributed energy business.

Versatile CHP powers modern microgrids. Fueled by natural gas or other fuels, and in a small form factor suitable for most site requirements, reciprocal engine or microturbine CHP is a highly efficient technology for generating electricity plus steam, hot water, or chilled water from its waste heat. Despite being one of the most mature DER segments, CHP still represents less than six percent of installed electricity generating capacity globally, according to Navigant Research. That appears about to change, however. Navigant sees growth in the industrial and commercial CHP market based on renewed interest in this proven technology platform across all regions (annual worldwide revenue from industrial CHP is expected to grow from $19.7 billion in 2013 to $29.8 billion in 2023) [94]. The small, modular form factor of microturbine CHP brings added versatility to the value equation: When urban space is at a premium, fitting adequate solar PV on a relatively small roof may prove infeasible, but installing a rack of microturbine CHPs in a basement is a great urban solution. Pair that with greater efficiency from putting waste heat to work and you have a winner. The fact that microturbine CHP units can be dispatched makes them natural complements to renewable wind and solar PV and an ideal fit in a modern microgrid. An interesting tale about CHP in New York, which lags other large states in solar PV penetration but is way ahead in adoption of CHP, can be found in [95].

Microgrids and nanogrids. Microgrids, covered in more detail below, have shifted in five years from concepts to reality. The idea of islanding load has become far more accepted than it was when this book was originally published, even if the task has not gotten much easier. A nanogrid emerges when the management scale of a microgrid gets smaller but still follows similar principles. Nanogrids may become a new operational strategy for building owners and managers. While Bruce Nordgren of Pacific Northwest Labs has been studying this concept for years, it appears that the time has come to consider the possible implications of such a concept: With nanogrids, management of the electricity could expand to include buildings, appliances, and local circuits that operate independently of the grid, or signal the grid when grid integration is indicated [96].

DRMS, DERMS, and ILM: The dawn of smart load. We celebrate the dawn of smart load with three acronyms—12 letters that must be some kind of record. A gentle revolution is under way in these technologies, portending a paradigm shift in grid operations. Over 100 years, we used arcane planning and watched the weather to make demand estimates, then regulated voltage from central power plants in increments to keep the grid in harmony, "chasing load," because we couldn't control it. These three technologies promise to change all that, bringing load awareness and sensory data to bear. DRMSs enable management of disaggregated, multivendor DR systems [97]. DERMSs represent an emerging category of demand side control systems led by Alstom (acquired by GE in 2014) and Siemens [98], and ILM is new software just becoming available that integrates DRMS and DERMS. Together, these technologies raise the prospect of a new paradigm emerging, which we might call "load follows resources." Smart load's day is dawning.

LEDs don't just save energy; they're a different light value proposition.   McKinsey's 2012 Global Lighting Market Model saw the LED share for general lighting going from 45 percent to 70 percent by 2020 [99]. In 2013, Wintergreen Research predicted that LED lighting global market would grow 45% annually from 2012 to 2019 [100]. Regardless of the study, it's clear that price drops in LEDs are starting to resemble the famous Moore's Law, now causing a paradigm shift in the lighting industry with very short product life cycles and innovation driving sales. Moreover, now that prices for LED light bulbs have dropped well south of $10, competitors are seeking means to differentiate, and one of them is to make LEDs smart [101]. When you can control an individual light bulb from your smart phone, we're in a new world—apparently one where lighting energy requirements have dropped permanently by more than half.

Trend 2: Edge Power Policy Trends

Rise of decentralization/DER.   When we introduced the advanced smart grid in Chapter One, we used trends in telecom and IT to illustrate the direction of change in the electric industry. We highlighted the rise of mobile applications, smaller form factors, and networking as logical consequences of Moore's law and Metcalfe's rule. As computing and networking grow more powerful and nimble—and less expensive—sensor devices proliferate, mobility is enabled, and transformation of traditional industries becomes more the norm than the exception. From the utility industry perspective, it was quite logical in the first decade of the new century to look at decentralized energy and talk of grid parity, the point when new, relatively expensive alternative energy technologies would stand up to traditional technologies without subsidization. For traditionalists, grid parity appeared to be years into the future. The grid parity argument continues based on perspectives; many continue to evaluate new energies primarily on $/MWh. Using measurements such as levelized cost of energy (LCOE) and levelized avoided cost of energy (LACE), predominate types of DER like PV, EVs, and ES may still appear to be years from becoming serious threats to existing fossil-fueled resources. However, that view is less confident than it was before significant price declines in renewable energy over the past few years [102].

Moving beyond low-cost energy.   An interesting development over the past two years has challenged the grid parity argument. Rather than evolutionary displacement based on commodity pricing under the LCOE framework, the rise of DER accelerated and became a force on its own, driven in part by cost reductions, but increasingly by value and innovation in three forms: choice, security, and control. An expanding array of energy choices has appeal to a world of energy consumers accustomed to monopoly grid power. As extreme storms caused more frequent and longer-lasting power outages, energy security rose as a consumer priority, differentiating non-grid DER from traditional grid power. Finally, greater control of energy conforms with the experience of consumers in other parts of the economy. Moving from grid power provided by monopolies to more controllable DER options offers consumers something the grid cannot—the ability to tailor personal energy to provide the value consumers seek, based upon their own conditions, just as they do in other parts of their lives. Not unlike the value offered to travelers by automobiles in the age of the railroad 100 years ago, utilities face an adaptation challenge. In the face of such value shifts, utilities must redefine themselves and their value proposition in terms of new competitive standards such as power continuity (lower exposure to outages), green power (a lighter carbon footprint), and tailored power (adding control elements), not to mention price stability (lower exposure to rate increases).

Proactive market restructuring. Regulators have tracked these disruptive trends and by varying degrees have adapted state rules to accommodate the coming transformation of both utility business models and electric industry restructuring. In the following paragraphs, for focus we limit our attention to recent activity in some of the most adaptive states, Hawaii, California, New York, and Massachusetts,

Hawaii. The Hawaii Public Utilities Commission (PUC) issued findings in several dockets in April 2014, sending a strong, public message to Hawaiian electric companies that they are not moving fast enough to lower utility rates and connect more solar PV systems and laying out an action plan on utility goals, including reducing energy costs, being more proactive to emerging integration challenges, and embracing customer DR programs. The PUC's guidance and direction focused on IRP [103], system reliability [104], DR [105],. and upcoming rate cases [106]. Hawaii electricity issues have historic, widespread attention, generally seen as leading indicators of the future of electric policy issues worldwide. The soaring electricity rate from the expensive cost of importing oil and natural gas and the prevalence of natural resources (good solar insolation and steady strong wind) combined with small distribution grids and deep-pocketed consumers present a perfect storm of operational challenges and transitional influences, driving Hawaii into the future at a more rapid pace than anywhere else in the world. Like California, Hawaii has emerged as a canary in the coal mine—a living laboratory for electric industry transformation.

California. California has a unique role among the 50 states—it is traditionally where the U.S. public looks for clues about the future. California's power utilities have great experience going first, often experiencing difficult growing pains as pioneers in a variety of areas, from EE 20 years ago, to retail power markets last decade, to smart metering and MDM, to smart grid, renewable energy, and ES. California leads in almost every category of innovation associated with electric utilities. The legislature and the PUC have driven utilities to be the first to support building EE, to deploy smart meters, to aggressively deploy wind and solar PV systems, to integrate storage, and to develop smart grid roadmaps. Trends seem to originate in California. So, with transformation, we will continue to look to California for an object lesson on how or how not to integrate these multiple trends into a new energy system.

New York. The N.Y. Public Service Commission (PSC) recently released an order, Case 14-M-0101—Reforming the Energy Vision (REV), which describes a new energy paradigm to incorporate market and technology changes and place DER on equal footing with grid-delivered commodity kilowatt hours, which it calls distribution service platform provider (DSP) [107]. Since its introduction in spring 2014, REV has gained national, even global attention as one of the first significant attempts by a state regulator in the United States to reimagine the power paradigm and acknowledge the challenges facing the industry. The electricity industry can be said to have begun in New York, the center of the pioneering work of Edison, Tesla, Westinghouse, and Insull. Edison's Pearl Street Station became the model of early DC generation to support central lighting districts, and Buffalo is the site of the first AC long-distance generation from a power plant in Niagara Falls to the city of Buffalo. The REV represents an attempt by Governor Cuomo's office and the N.Y. PSC to put New York in the front of the line to lead the development of a new energy paradigm. N.Y. PSC chair Audrey Zibelman, a former CEO of Viridity and a leading electric industry figure associated with innovation and new energy concepts, is crafting a model to structure a market platform that encourages integration of DER but leaves specifics to the marketplace [108]. N.Y. utilities have a ringside seat to novel energy business planning. Con Ed is a world-leading electric utility feeding Manhattan and New York City, and other large N.Y. utilities include National Grid, PSEG Long Island, Central Hudson, Orange & Rockland Utilities, NYSEG, and Rochester Gas & Electric, as well as a number of smaller municipally owned distribution systems.

Massachusetts. With the exception of a smart grid pilot by National Grid, this state would have to be considered a late adopter of smart grid. National Grid began a small smart grid pilot involving 5,000 meters in Worcester, MA, and has since expanded it to include 15,000 customers to evaluate integrating home automation, dynamic pricing, and DA. However, the MA Department of Public Utilities (DPU) is making up for lost time, having decided in 2013 to aggressively accelerate the deployment of smart grid technologies in the state, with an order mandating a smart grid roadmap among investor-owned utilities [109]. The MA DPU is requiring its IOUs to develop a grid modernization plan (GMP), which is required in the first phase to include a comprehensive plan for advanced metering (CAMP), with a three-year mandate to complete an AMI rollout. Additional requirements will include TOU rates and plans for EVs and EV charging.

Texas. Texas is a unique territory to watch, but not because of any current regulatory initiative, as above. Years ago, the Texas legislature approved power deregulation and more recently, smart meters and a massive transmission line project called Competitive Renewable Energy Zone (CREZ), all measures designed to stimulate a market-first approach in this market friendly state. Alone among the 50 states with an isolated statewide operating area [Electric Reliability Council of Texas (ERCOT) is connected to the Eastern and Western grids only through DC ties], Texas stands out in other categories. Of all the deregulated states, Texas has the most robust competition [more than 50 retail energy providers (REPs)]. Its competitive retail power market is ferocious and innovative—some Texas REPs offer free nights and weekend plans with their power contracts. Texas has one of the most robust wind energy markets, with installed wind energy resources representing 15% of its generation mix, expected to grow to 20% by 2020. Solar PV insolation is superb in Texas, and Texas has a large, growing population hungry for power, offering solid potential for a robust solar PV market to develop. (Notably, Texas is one of only seven states without a net metering policy, and Texas is only sixth in installed solar PV capacity despite such natural competitive advantages.) To date, a large part of installed solar PV capacity in Texas can be attributed to innovative city-owned utilities in Austin and San Antonio (see Austin and Solar PV below). Furthermore, Texas is home to an historic boom market in oil and natural gas, providing added promise for natural gas-fueled DER. Unique legislation approved in September 2013 (HB 2049) allows property owners in competitive energy markets to install and operate DER in their buildings as an unregulated utility to provide tenants power, cooling, and heating—a great opportunity for natural gas-fueled CHP, and clearly a game-changing new law for a state that seems poised to break out as a DER leader.

Germany and feed-in tariffs (FITs). Germany made a far-reaching policy decision to transition from fossil fuels to renewable energy, made possible by the strong nuclear and fossil fuel base load from neighboring nations that maintain the frequency of the European grid. Accordingly, Germany deserves the gratitude of all solar owners thanks to its pioneer efforts to make PV and DSPV commonplace. Using FITs, Germany made solar panels a matter of national energy policy. Heavily influenced by the Green Party, the German government obligated utilities to pay a guaranteed high price for the output of solar panels, and investors raced to cover available rooftops and empty lots with PV, securing a stream of revenue for years to come. Few can dispute the success of early market support to drive adoption of solar PV, as economies of scale from FITs drove worldwide solar PV steadily downward. However, such FITs were intertwined with the complex energy ecosystem: Retail energy prices in Germany steadily climbed to support utility FIT payments to PV owners, and large power plants were displaced by DG. German utility giants E.ON and RE were early indicators of these stresses, shuttering a relatively new power plant (E.ON) in spring 2013 [110] and revising the utility business model to shift to energy services (RWE) later that year [111].

Germany and nuclear, coal generation. A parallel to Germany's push for renewable energy (i.e., Energywiende) was the German government's policy decision in 2011 to shutter all 27 nuclear energy plants by 2022 in the wake of the

Fukushima disaster. A perverse unintended consequence became apparent from 2012 onward: Lignite coal-fired power plants fired up, coal's portion of the national energy mix climbed to 47%, and greenhouse gas production actually went up in Germany [112]. Since then, a lively debate has continued over the pace of change and the choices made, both to promote renewables and scale back nuclear, all in hopes of addressing climate change with minimal economic impact. Can the equation be balanced? What will restructuring look like over time?

United States and coal generation.   The U.S. EPA's Clean Power Plan draft order released in June 2014 focuses on existing power plants, providing options to state governments to steadily move the nation's central generation portfolio toward less carbon impact, providing multiple options with a clear objective: to reduce pollution from old coal power plants. A recent Forbes article noted a natural shift toward natural gas and away from coal already under way [113]. Such change will come slowly, at least from the government's perspective, with state plans required by 2016 and implementation not in effect until 2020. The net-net of these new rules will be to favor renewable energy, DER, natural gas, and nuclear over coal; the rule will require expensive carbon sequestration retrofitting of older plants, which many believe will leave mostly the most modern, cleanest coal plants already in place to stay open after the new rules go into effect. Older plants will face expensive retrofits, and few new coal plants are expected to be built. Utility expert Bob Bellemare weighs in with a counter view in a 2014 Fortnightly Spark article, warning observers not to dismiss coal plants and carbon sequestration too quickly, while acknowledging the momentum this order gives to cleaner alternatives to coal [114].

Japan and Fukushima.   In Japan, the nuclear disaster at Fukushima proved both a shocking national tragedy and a long-term challenge for a nation heavily dependent on reliable, affordable grid power. A March 2013 article noted that of 52 nuclear plants operating before the disaster, only two continued after [115]. In the face of such drastic change, Japanese utilities and consumers had to reexamine the potential of energy conservation (DR) and DER to take a greater role in meeting power needs, coordinated by a smarter grid. Policy shifts included a smart meter mandate to enable the grid and aggressive DR goals, as well as DER goals supported by FITs, with a 2020 PV target of 28 GW (53 GW by 2030) and around 80% coming from DSPV. Consider that U.S. solar capacity remains in the single digits measured in gigawatts. Japan's countrywide RPS goal of 20% by 2020 is certainly aggressive. Life must go on, as Japan has showed us, and the loss of nuclear has driven tough choices and proved a boon for alternative power sources.

Understanding Germany and Japan.   The debate rages in late 2014 about the meaning behind these events in Germany and Japan, as noted in a late September article [116]. It is not so straightforward that a massive amount of renewable energy on the grid causes reliability problems or that the removal of nuclear has spurred a demand for coal. There are elements of truth to both of these positions, but reality is more nuanced. Similarly, ES is not necessarily a panacea to grid operating challenges. A report released by the Brookings

Institution titled Transforming the Electricity Portfolio [117] addresses these and other issues highlighted by the unique positions in which Germany and Japan find themselves, out on the frontier of infrastructure and business transformation. Stay tuned.

The United States and power purchase agreements (PPAs), leases. In contrast to top-down mandates in Germany and Japan (and China, below), solar PV adoption in the United States has so far been based on the synergy of government stimulus and private sector business innovation, which promotes third-party ownership and new forms of leasing over equipment ownership. A September 2013 article notes that the introduction of the long-term PPA, together with price drops, dramatically accelerated the deployment of PV systems by lowering risk and increasing value [118]. Rather than selling solar PV systems to a small group of privileged buyers leveraging rebates, innovative companies like Solar City, Vivint, and Sun Run went after much broader markets with no-money-down contracts that sold the output of the systems in long-term contracts, obligating signers to host rather than own the system on their roof and purchase its output with regular monthly payments. The PPA offered energy consumers a partial independence that proved irresistible as it became more widely adopted.

China and PPAs. China remains a fascinating study in the evolution of the power industry, as its seemingly insatiable hunger for power to support its maturing economy forces innovation and creativity on all fronts. In September 2014, Chinese leaders announced policy support for decentralized energy as a fundamental strategy [119]. The country's aim is to add 8 GW of DSPV and 6 GW of utility-scale PV by the end of 2014. Significantly, this is at a 7 cents/kWh tariff. Furthermore, meeting that target would put China firmly in the lead globally as the year ends. China will have gone from 800 MW of DSPV to 8 GW—a tenfold increase in only one year.

Central America and PV capacity. A September 2014 article trumpets the dramatic potential that solar PV has in Central American countries, projecting a 1.5 GW capacity goal by 2018 among six Central American countries [120]. Demand growth is expected to drive deployment of both utility-scale and DSPV systems as costs continue to fall. In a region largely dependent on hydropower, but also punctuated by frequent grid outages, it is expected that utility PV PPAs will help diversify the region's energy resource capacity, even as DSPV projects proliferate in complementary fashion.

Canada and utility of the future. New Brunswick Power (NBP) is a small (330K meters), vertically integrated crown corporation (owned by the provincial government). This type of public/member ownership in Canada resembles a blend of a large municipally owned utility (MOU)/coop in the United States. With an ambitious vision, but faced with significant strategic planning challenges, NBP engaged Siemens in 2011 to do strategic planning, and in 2012 negotiated a 10-year service contract to realize its "utility of the future" vision. A core component of this vision is to place a growing emphasis on demand side resources, leveraging more fine control of DER to replace energy resources normally found in central power plants, thereby avoiding significant capital expenses associated with new power plants in the medium and long term. This long-term partnership for business transformation is a unique example of early-adopter commitment to change.

California energy policy: DR and DER.   When a court ruling challenged FERC Order 745 as unconstitutional, essentially constituting federal overreach of state duties, the DR world faced an unprecedented challenge. The California legislature stepped in with CA SB 1414, a fix that directs utilities to use DR first to meet power needs [121]. Notably, DR is not seen as an occasional emergency backup to central power plant supply, its traditional role. Now, California has turned the tables, elevating DR to primary supply source—the demand side has suddenly become far more prominent in resource planning. While there remain mountainous challenges to get consumers to buy into DR, this line of thinking sees high growth potential for DR as a strategic resource. DR will become a key mechanism as well to shift consumption to time periods when DG and ES are abundant, enabling far more DER onto the grid over the coming years.

California energy policy: ES.   CA AB2514 directed the CPUC to develop rules setting a statewide goal for ES penetration of 1,325 MW of new ES by 2014, an amount sufficient to supply one million homes with stored electricity at peak use [122]. The new rules also require utilities to issue requests for ES every two years, under three principal uses: (1) transmission interconnect; (2) distribution interconnect; and (3) behind-the-meter uses.

Hawaii energy policy: renewable energy.   As HECO rewrites its IRP under direction of the Hawaii PUC, the 2030 goal for RE has been more than doubled, from 35% to 65% of total resource supply [123]. RE is at 18% now, under the RPS of 35% by 2020. PV is the majority component of RPS currently. With this IRP, HECO is trying to lower the price of the customer bill by substituting PV for traditional fossil fuel generation, which depends on expensive imported fuel. So we see in both Austin and Hawaii, as solar PV and DSPV moves down in price and becomes more accessible, RPS goals accelerate to drive even more penetration. Even with more robust goals, the challenge remains on how the grid will handle such a dramatic expansion (see Nessie Curve below).

Hawaii energy policy: ES.   As we go to press, HECO is engaged in the purchase of 200 MW of ES, seeking to integrate grid-scale ES to address the challenges of RE integration—principally, intermittency from variable wind and solar PV production [124] (again, see Nessie Curve below). HECO seeks 30-minute increments from ES projects of at least 60 MW in size, collectively 200 MW of capacity. With California and New York, Hawaii is at the forefront of integrating ES into its grid.

New York and ES.   ConEd is getting into the ES game in a big way, with its aggressive push to use DR and DER to reduce or shift load to enhance grid resiliency but also to avoid large investments in traditional grid infrastructure [125]. As ConEd looks to shutter its 2-GW Indian Point nuclear facility, it is evaluating the potential of load reduction, together with NYSERDA. Notably, the comparison has been made with California's self-generation incentive program, which supports DER growth with one-time, upfront rebates on equipment installed behind the meter.

Duke and solar PV.   In September 2014, Duke announced its intention to seek PPAs totaling 128 MW of utility-scale solar for a total investment of $500M, adding to its current 150 MW of solar capacity [126]. Duke is seeking to meet the relatively conservative North Carolina RPS of 6% by 2015.

Austin and solar PV.   On the heels of a recent PV PPA that came in at about 4.5 cents/kWh, the city council in Austin went all in on solar by passing an ordinance with an aggressive set of goals for PV and storage on August 28, 2014 [127]. First, on the PV front, Austin established twin goals for utility-scale (centralized)—600 MW by 2017 and for DSPV (distributed)—200 MW of DSPV by 2020. For ES, the ordinance sets a target of 200 MW , which will likely be needed for Austin Energy to continue to manage grid reliability. With Austin Energy's peak capacity at about 2,700 MW and renewable penetration already at 35% (945 MW), adding 800 MW more would push renewable energy penetration to near Denmark standards, somewhere north of 60%...in roughly six short years. If this trend were to get legs, we may be able to start talking about a tipping point for new edge power utilities. Not so fast, however. As we go to press with this second edition, Austin Energy management has pushed back on the city council's ordinance, adding a cost figure to these ambitious goals—$1B over 10 years—and challenging a complete fossil fuel withdrawal, recommending the addition of a new gas plant [128]. This measure puts the authors of this book on the front lines of this transformation, as we watch this aggressive move to make solar PV the default energy source.

City versus IOU: the Boulder-Xcel Energy saga continues.   In 2012, Boulder held an historic vote and by a slim margin, decided to approve a study of the feasibility of exiting the Xcel Energy system, setting in motion a contentious process to create a municipal utility and gain autonomy over its energy future. The city's goals include getting to 40% renewable as soon as possible and being completely carbon free by 2050. This conflict puts the reorganization of the U.S. power industry under the microscope, at least from the municipalization perspective. While Xcel Energy is required to balance regional needs under the guidance of the Colorado PUC, Boulder and cities like it seek greater independence and autonomy [129]. As the saying goes, "Breaking up is hard to do."

Going full renewable.   In September 2014, the small municipal utility serving Burlington, Vermont, announced that it had achieved its goal of 100% renewable energy sourcing with its purchase of a small local hydroelectric plant [130]. Wind, water, and biomass support this university town now, offering a lesson for fellow Vermonters. Vermont has a 90% RPS goal by 2050.

SMUD and residential PV.   So, what happens when a utility goes still further to find a way to stimulate residential solar (DSPV) penetration? Perennial utility pioneer SMUD is showing us the way to think about customers instead of ratepayers [131]. Already a leader in rooftop solar penetration (growing from 1,000 rooftops in 2012 to over 2,000 in 2013), SMUD faces daily calls from Sacramento consumers about PV, developing into its new role as a trusted energy advisor. SMUD is using data analytics to support its customer portal operations. What's next? Partnering with third-party PV companies will provide a direct-to-customer solar service, using streamlining efficiencies to support still further growth in DER penetration.

DER and strategic rate reform.   Rocky Mountain Institute (RMI), a thought leader on new energy, published a report in August 2014 titled "Rate Design for the Distribution Edge," highlighting yet another trend: using rates to integrate and promote DER [132]. With more sophisticated rates, utilities can get to the underlying value of pricing to address unintended consequences like cross-subsidies. Focusing on three key areas, RMI introduces a path away from basic ratemaking: (1) unbundled pricing based on attributes of value; (2) temporal granularity that addresses the time component of rates (peak versus off-peak); and (3) location granularity that acknowledges that rates can be used to adjust to localized congestion that raises the costs of distribution (and the value of DER at that spot). RMI is shining the light on a key differentiator for utilities to remain competitive (i.e., new strategic rates), a way for utilities to address the disruptive operational impacts of DER and a way to use DER to improve grid operations.

Net metering reform.   In the early days of DSPV, utilities encountered two key challenges. First, they faced the safety risk to line workers when DSPV systems kept a feeder line energized during an outage. Second, they faced the market challenge of balancing compensation for DSPV owners for production beyond their own needs (when electrons flow out onto the distribution feeder) and the costs to utilities of adding PV to the system. The solution came in the form of net meters that acted as both breakers to address safety risks and an economic transaction tool to track production and grid transfer. However, the rates that utilities paid DSPV owners for power varied significantly. Net metering reform addresses the debate between utilities that seek to cover all the costs of adding DG, challenging negative operational impacts and cross-subsidies among rate-payers [133], and PV proponents who seek to receive full payment for the value of their excess production and to stimulate greater deployments and market penetration [134].

Value of solar tariff (VOST).   On the far end of the reform scale is the VOST developed at Austin Energy in October 2012 and picked up in 2014 by the state of Minnesota. The VOST offers an alternative to net metering, at least from an economic standpoint, by compensating DSPV owners based on the full value of their production and utility costs using a complicated tariff formula. Going beyond the minimum of avoided fuel costs to include such value as avoided congestion and avoided capital expenses, VOSTs represent a debatable alternative to net metering, even described by some as a type of FIT, albeit a somewhat flawed solution still subject to annual variations that discourage long-term investors [135]. Notably, this economic debate leaves out the frustration of DSPV owners during grid outages, when the PV system is left inoperable because of its reliance on a grid signal. (See PV plus storage below, and for a fuller dive into these topics, see Chapter 7's discussion of positive power, DSOs, and pricing.)

Supply versus demand, old versus new. As DER is increasingly accepted as a substitute for traditional cenral grid resources, it becomes necessary to reexamine the way that generation, transmission, and distribution investment planning is conducted for both assets. According to [136], the shift from a focus on supply planning to embrace demand-side planning as well is needed, or the risk of stranded investment rises dramatically. Conservative system planning involves projecting system needs far into the future, as much as three decades, to assess system investment requirements. However, while DER is increasingly able to address the needs that traditional central generation and T&D investments do, DER is still not embraced fully by planners. This oversight highlights a growing risk and the need for an adjustment, trading old ways for new ways. Ironically, conservative risk-adverse behavior may prove more risky than progressive behavior in the face of such change.

Peak management. All load is not created equal, and the same must be said for resources. As more and more DER is added to the grid, the sensitivities associated with meeting peak load have been revealed as a key challenge for grid operators, as noted in two interesting curves in circulation in California and Hawaii, detailed below. The challenge of meeting peak demand in a new more complex operating environment showcases the need for dispatchable resources in a centrally managed grid and real intermittency constraints with more and more DER.

Duck curve. Understanding the duck curve may be seen as a proxy to understanding many of the issues outlined in this section, as well as the potential of new approaches. Regional increases in DER displace traditional central generation (i.e, the body of the duck), but as the sun goes down and DER production wanes, the need for more power rapidly returns (the neck of the duck). A May 2014 article outlines 10 ways [137] to solve the challenges associated with the duck curve—note the balance between storage and demand changes: (1) use EE first to address the evening demand ramping—LEDs offer a significant opportunity to address this evening challenge; (2) shift the orientation of PV panels more to the west to extend production into the peak time (an approach that has been confirmed at the Pecan Street testing site in Austin); (3) add storage to solar, whether by using some concentrated solar PV or adding batteries to DSPV, to address the evening ramp; (4) gain distributed thermal storage by networking electric hot water heaters; (5) leverage HVAC capacity with ADR or thermal storage like ICE energy to address the ramp; (6) retire old power plants that are too inflexible to deal with this new challenge; (7) use "concentrated" demand charges during peak hours to induce changes in load behavior; (8) leverage the modular aspect of DER by matching deployment in areas where the ramping challenge is most acute; (9) ask more of DR programs—we have a long way to go to develop DR as a strategic resource; and 10) leverage regional differences and the transmission system—for California, that means more active management with the Pacific Northwest. The optimistic tone of this approach doesn't make light of the challenges associated with the duck curve but highlights innovation and leverage of all the new tools at our disposal.

Nessie curve.   Not long after the duck curve was introduced to describe upcoming grid management challenges in California, a variation, the Nessie curve, was introduced at Distributech in San Antonio in January 2014 [138]. Why Nessie? The longer neck of ramping in Hawaii's existing load shape more closely calls to mind the elusive creature from Scotland than a duck. With Nessie, HECO planners highlighted current challenges that needed their immediate attention. The operating challenge is more acute and immediate in Hawaii due to higher grid prices, smaller grid operating leeway, and accelerated DER penetration. HECO faces three unique challenges: backfeeding, which happens when DER production exceeds demand (when Nessie's belly dips below the zero axis), exposing sensitive grid equipment to reverse voltage flows; it also faces limited system awareness, as many, many more sensors are needed to meet the granular information feedback required to manage operational threats proactively; and unpredictable continued power need to meet peak demand when solar quits producing, driving the need for ES and/or peak load management (the long neck of Nessie). Needless to say, the situation in Hawaii has drawn considerable attention and curiosity, as this unique future-shock situation unfolds.

Hurricane Sandy and system resiliency.   As Hurricane Sandy moved up the coast and slowly built up power, the words "perfect storm" began to be heard. This mammoth storm struck at the most densely populated point on the Atlantic coast of the United States, unleashing incredible force and laying bare the vulnerability of our vital electric grid. ConEd, PSEG, and other utilities responded bravely to restore power. However, buried distribution lines and underground distribution substations could not avoid the floodwaters of the storm surge, and restoration was hampered as workers actually disassembled equipment so it could be dried out before going back into operation—electricity and water do not mix well. As restoration proceeded, stories about the drama during the outage abounded, and lawsuits began to be filed amid finger-pointing to assign blame. Beyond the drama and recovering from huge economic costs associated with property damage, this storm put the electricity industry under the microscope. Calls for increased system resiliency echoed throughout the Northeast. A year later, in late October 2013, the DOE outlined its continued commitment to improve system resiliency, focusing on grid modernization, outage recovery process improvement, and redundant power systems [139].

Microgrids and system resiliency.   Microgrids, from Maryland to Princeton to Manhattan, kept supplying power during the massive grid outages of Hurricane Sandy, burnishing their reputation as a strategy for power continuity [140]. The Hudson Yards microgrid project in the planning stages received media attention in August 2014 [141]—the $20B project planned for midtown Manhattan will be the most ambitious, expensive urban project ever, and it will leverage what we might label the new tools of sustainable smart cities: (1) data-driven processes facilitated by modern communications to improve quality of life and lower costs; (2) EE to limit energy requirements and costs; (3) automated, intelligent waste disposal through pneumatic tubes to eliminate garbage trucks; and (4) microgrid technologies to localize energy production and consumption, ensure power continuity, and support system resiliency.

Trend 3: Business Model Trends

The utility death spiral.   As they say in 12-step circles, the first step in dealing with a problem is to accept that there is a problem. We evaluated DER and its potential impacts on the electric utility business model in the Pecan Street Project in 2009 (see Chapter 5), and with The Advanced Smart Grid, written in 2010; furthermore, we anticipated the challenges inherent in the steady march of edge power, going so far as to subtitle our book, Edge Power Driving Sustainability. We see this meme entering the mainstream in late 2012 and early 2013 with the publishing of two key reports: first, the Reznick Institute published Grid 2020 in September 2012 [142], dissecting the challenges presented to the grid by renewable energy and DER. Then the EEI published Disruptive Challenges, a white paper that acknowledged what many had begun talking about openly, that is, the real and present threat to the status quo of the traditional retail electric business model posed by the accelerated maturity of DER [143]. Now, the cat is fully out of the bag, as the remainder of this section shows.

Inertia, paralysis, inhibitions, and rigidities.   A June 2013 study presented at a conference in Barcelona is one of the best third-party commentaries on the size and scope of the challenge facing electric utility incumbents [144]. With a close look at how disruptive technological innovations might affect electricity distribution network utilities, this study goes on to identify several factors that will inhibit incumbent efforts to respond to a building crisis. With a group of 18 electricity distribution regulated monopoly electric utilities, this study makes three fundamental conclusions: (1) disruptive DER technologies "are likely to trigger the 'creative destruction' of existing natural monopolies and render incumbent business models unsustainable"; (2) adapting to disruptive technologies is made still more difficult by having to transition from monopolies to competition while dealing with that disruption; and (3) incumbents will need to overcome the inertia and paralysis inherent in their current operating models and organizational structures to avoid succumbing to a "death spiral scenario."

Wall Street weighs in.   This open shop discussion inside the utility industry was soon joined by a variety of financial journals and media intent to weigh in with their own outsider perspectives. The message of increased risk was a common thread. Three studies in particular stand out (though this is a mere sample of the literature in 2013–2014). UBS released a report in December 2013 describing the "difficult road ahead" for utilities globally [145]. The report provides a valuable analysis from the global perspective, citing the plight of utilities in Europe since 2008, noting that the top 20 European utilities have lost half their value based on a perfect storm of conditions in deregulated markets: falling demand for electricity, too many fossil-fuel plants, and declining wholesale costs for renewable energy, citing EE and solar PV as primary causes. Then, Goldman Sachs provided its perspective, describing the seismic shift in focus and attention. This report highlights the dramatic drop in prices for PV and storage, anticipating widespread LCOE-based grid parity in 2033 [146]. Soon thereafter, Citigroup shared its views, declaring a new "age of renewables,". citing the growing stability of renewable energy cost declines in comparison to pricing volatility with natural gas and long-term challenges with coal and nuclear energy [147].

The economics of grid defection.   Perhaps the best is indeed the last, or the latest word on disruptive impacts and new business models. The folks at RMI published an excellent report available for download on their site on the subject of grid defection [148]. Calling the combination of DSPV and ES a "utility in a box," RMI draw the line of distinction. When a true alternative to grid connectivity arrives and becomes affordable, it will present a challenge to the existing utility business model, opening the door to those who wish to leave the 100-year old social compact, to "defect" from the grid. RMI predicts that it will be about 10 years before the inevitable compounding behavior really takes hold. With this and other discussion above, the question becomes, "How will utilities react with new business models?" Moreover, what will transformation look like?

Surviving the coming utility revolution.   Smart Biz hosted a webinar with this provocative title in September 2014 [149]. Unique in this webinar is a detailed discussion from prominent commercial customers on their particular perspectives on change and what they believe utilities should do to take action now. Wal-Mart's David Ozment shares his wisdom from years contemplating these trends, concluding that utilities should focus on creating real value with new business models coinvented with customers, centered on reducing costs and increasing sustainability. Microsoft's Brian Janous described DER as part of Microsoft's core infrastructure, envisioning an integrated grid that leverages the best of centralized and DER. The bottom line for these large commercial entities is that they are focused on their core businesses and expect to work with utilities going forward, not to replace them.

Grid as backup power provider.   In July 2014, Microsoft announced a milestone in its corporate strategy for powering data centers with collocated power plants, describing a 175-MW PPA for wind energy over 20 years with the Chicago Pilot Hill Wind Project [150]. This amount of energy, equivalent to what is needed to power 70,000 residences, highlights a trend of securing clean, nonutility energy for critical power needs to lower long-term costs and make a corporate energy portfolio ever more green.

Transportation as a service.   In a move that might be perceived as piecing together software components to become a Kayak equivalent for on-the-ground mobility, Daimler bought two more transportation related app companies in early September 2014 [151]. Adding to its current portfolio (car2go, a car sharing app, and Park2gether, a parking space finder), and in order to compete with Uber, Daimler Chrysler purchased RideScout—to find all forms of rides, public and private—and mytaxi—to track the service chain of a taxi ride. With this growing array of software, a vision of transportation-as-a-service is emerging.

This is a disruptive trend of using the "as a service" business model to "self-disrupt," or change from within.

EV charging stations stumble. Five years ago, BetterPlace gained widespread attention for its ambitious plans to build charging networks in Israel, Denmark, and Hawaii, but it ultimately succumbed to bankruptcy. Likewise, ECOtality, a recipient of $100 million in DOE funding, built out its Blink charging network to more than 12,000 charging stations, only to go bankrupt and see its assets purchased in October 2013 by Car Charging Group, a Miami-based company [152]. These very public early failures have sparked a debate over whether charging stations should remain proprietary or would be better off being part of an open network using open standards [i.e., open charge point protocol (OCPP), now available in over 50 countries] [153].

Microgrid becoming a new standard. Perennial utility disrupter NRG has done it again, cutting a deal with Vermont's Green Mountain Power in early September 2014 to build a community microgrid for Rutland, Vermont, to be jointly owned and operated [154]. Should NRG gain more traction as a microgrid developer partner for utilities, a new more efficient business model for electricity production and distribution will have arrived. A few weeks later, ComEd received a DOE grant to pursue its vision of a network of linked microgrids [155]. The $1.2M will support development of a controller to be used to manage clusters of networked microgrids—a new business model for utilities if ever there was one. ComEd will collaborate with a group of science and technology partners including Alstom, Argonne National Laboratory, Illinois Institute of Technology, Microsoft, OSIsoft, Quanta Technologies, S&C Electric, Schneider Electric, and University of Denver.

DSPV and LEDs: modular, off-grid solar. Navigant Research predicts that the global annual market for solar PV consumer products will grow from $550.5 million in 2014 to $2.4 billion in 2024. Pico solar systems combined with low-energy lighting solutions are providing exciting alternative opportunities to traditional lighting (e.g.,. kerosene wick lamps and hurricane lamps) that offers to significantly impact pollution and public health [156]. With dramatic payback periods as low as one month [157], these new technology systems also significantly reduce carbon footprint. Ideally situated for areas where traditional grids are unreliable or nonexistent, these off-grid solutions are likely to easily adapt to the developed world as consumer devices.

Community energy: ILSR and CADER. These two agencies provide a community perspective to DER. The Institute for Local Self-Reliance (ILSR) [158] is a tremendous Web resource for the community energy perspective, asking the questions and providing the answers associated with new approaches to energy and business models. As a champion of localism and regionalism, ILSR is a natural champion of decentralized solutions. Similarly, Communities for Advanced DER (CADER) [159] is a group focused on promotion of DER solutions for communities using business models that support sustainable community development.

Transactive energy comes onto the scene.   A new business model offers an alternative to the retail sale of energy under regulatory oversight or in competitive retail markets. Dubbed transactive energy, this idea closely resembles the peer-to-peer energy trading concept we introduced as part of the SG3 conceptual overview in Chapter 7 of the first edition of The Advanced Smart Grid (see this edition's Chapter 7 for more on transactive energy). Transactive energy offers a market solution to enable billions of small edge energy transactions completed using the internet and software, a twenty-first century solution to be sure. Discussion on transactive energy grew rapidly after a LinkedIn group of the same name launched in early January 2013. By September 2013, the Gridwise Transactive Energy Framework was released at the SGIP conference [160].

A VPP in eastern Canada.   VPPs tie together diverse resources to act as substitute for a peaking power plant, thus, the name. PowerShift Atlantic [161], a compelling project in Atlantic Canada, has received worldwide attention for its ambitions to understand how aggregated loads can be used to offset the variance associated with onshore wind energy in New Brunswick, Nova Scotia, and Prince Edward Island. A team at New Brunswick Power has experimented with specialized equipment that stores heat in bricks and with distributed water heaters to demonstrate the management of demand-side resources to effect peak, whether it is curtailing demand or soaking up excess power.

Storage plus energy management: VPP as a service?   Sharp introduced its SmartStorage Energy Management Solution in July 2014 [162]. As advertised, it combines storage with energy management to create a service that helps its customers reduce or avoid demand charges. An option is to add solar PV to the system, which has it looking a lot like a VPP. This creative use of capabilities and price assurance associated with the corporate brand should encourage businesses to take more aggressive control of their energy expenses.

Renewable Energy (RE)/DER kitchen sink: holding company.   Perhaps DER's time has really arrived, if the actions of a Hong Kong property tycoon putting his money where his mouth is can be believed. Zheng Jianming, chairman and founder of Asia Pacific Resources Development Investment Ltd., may take public the holding company he created to manage a growing collection of DER businesses; the list of holdings includes battery technology, electric cars, geothermal systems and even units that use seawater to store electricity. Zheng was quoted in an interview in New York: "If a city were to implement all of these technologies it would basically be low-carbon. My vision for this company isn't just for China. I want to create a global company [163]."

How knowledge becomes money: making innovation a core competency.   Businesses that seek growth in the twenty-first century economy understand the connection between innovation and value. More value comes from finding clever ways to reengineer existing solutions and to use new technologies to create entirely new solutions, addressing heretofore unsolvable problems. Yet innovation remains a conundrum for electric utilities where the organizational culture and business processes are aligned to consistently achieve their commitment to deliver lowest cost power with high reliability. A July 2014 article in Forbes interviewed executives from Siemens, Fireman's Fund, and Applied Materials on the connection between innovation and value. According to Siemens U.S. CEO Eric Siegel: "The way Siemens thinks about creating economic value is through successful innovations. This relentless dedication to invention and innovation is what sets great companies apart. And we teach if you don't have innovation, you die...At Siemens, we believe our success is determined by our ability to anticipate and engineer the future. Rather than viewing ourselves as "business to business" or "business to consumer,",= we instead view ourselves as "business to society". As we look to engineer for today's challenges, megatrends like urbanization, demographic change, climate change, globalization, and digitalization continue to act as game-changing forces, creating new markets and new challenges [164]."

Analysis and Conclusion

After reviewing technology trends, policy trends, and business model trends, what do we find in terms of progress for the advanced smart grid, and what conclusions can we draw about our current experiences before going on to talk about what lies ahead in Chapter 7?

Before answering these questions, a general observation is in order: In the first edition of this book, Chapter 6 spoke mainly about beginnings, about mobilizing activity and shining the light on the ARRA grant stimulus, utilities, institutions and institutional activity. In the exciting days of the first edition's publication, "Smart Grid Has Arrived!" could have been our shouted headline. In this edition—while we decided to keep the important background material at the start of this chapter to describe the "foundation" of today's smart grid—however, we are describing an altogether different kind of excitement. As we do a massive rewrite of this chapter for this second edition, "DER Has Arrived" is the new headline, and this news begs many important questions, not the least of which is, "Okay, so now what do we do?" This chapter is a shorthand attempt to gain a better understanding of what has happened all around us over the past few years—quite a challenge in 20 pages or so—and to discern some meaning and insight on what may be done about it.

First, with technology trends, we see a variety of DER technologies maturing more rapidly than was generally expected (e.g., DSPVs, LEDs, and microgrids), and we see creative combinations of DER technologies bundled to craft solutions that more closely align with market needs and opportunities (i.e., PV plus storage, EV plus storage, EV/PV/storage, VPPs, and microgrids). The most remarkable growth has occurred with PV and LEDs, pushing potential paradigm shifts in their respective market spaces. Perhaps in part because LEDs apply only tangentially to the grid, a paradigm shift in lighting appears well under way—out with incandescent light bulbs and CFLs, too; in with LEDs and smart LEDs. With PVs, however, it's not so clean, and mountainous issues remain: notably, NEM/VOS/FITs, integration of intermittent resources on the grid, and impacts on the utility business model. The rise of PVs continues to dredge up a host of issues that demand our attention.

This leads us to the second category, policy trends, where we documented considerable evidence of restructuring, over 30 items. As we surveyed the landscape, it was hard to stop gathering news before going to press; there are so many stories to tell and always one more anecdote to add. We see attempts to address PV acceleration via both comprehensive and incremental restructuring in the markets most affected by early adopter activity. This puts cities, states, and entire countries voluntarily or involuntarily out on the frontier of change, offering the world abundant and profound lessons to guide emerging energy policy. In the United States, Texas, California, and New York stand out for strong leadership, but we really do have a laboratory here in the 50 states, with diverse approaches to change. Outside the U.S., activity in Germany, the U.K., Italy, Spain, Australia, and Japan is out in front, and experimentation is accelerating in countries around the globe. In the meantime, however, as the global energy transformation gets under way, most regulators, politicians, and leaders still lag behind. Accordingly, while the few pioneering policy leaders try to get out in front of the waves of technology change, it is innovation and invention that are running faster and extending farther than policy is willing to go. The world of electricity is inherently conservative, and that remains its main challenge in this dynamic environment, as numerous innovators outside the industry race ahead for a piece of the multi-trillion-dollar electricity pie.

With regard to business model trends, we see a rising awareness of the challenges ahead from both inside and outside the industry. The phrase *utility death spiral* has entered our popular lexicon—not as a foregone conclusion, but more as a warning. With debate now swirling around the true nature of the threat and what to do about it, it is time to acknowledge that the rise of DER is unprecedented for an industry founded on such core principles as system reliability, long-term investment in capital intensive assets for central power generation and electricity distribution infrastructure, and monopoly grants based on regulated oversight of rates, investments, profits, and balanced stakeholder interests. Our review of news items shows an incumbent industry reacting slowly to this threat, and an emerging alternative power industry based on DER technologies innovating around challenges and morphing rapidly to gain footholds. As this age old struggle of established giants versus pipsqueak upstarts plays out, we must realize that we have heard this story before and we know roughly how it turns out. Generally, the future is not bright for those incumbents unwilling or, in the end, unable to adapt. Consulting firm CMG [165] makes the case for change succinctly in its 2014 whitepaper *Disruption Becomes Evolution: Creating the Value-Based Utility* [166]:

> The "death spiral" for utilities represents one potential scenario, but it is not as inevitable as some would like us to believe. That said, the threats are real, and the utility industry must act now because:
>
> · Utilities are historically conservative and risk-averse. They need to learn to accept more business risks, leverage new technologies earlier, and take aggressive steps to alter their business model and improve their customer relationships.
>
> · Utilities are at risk, yet if they adapt intelligently, they can become more viable than ever before with the right strategic plan, roadmap, and technology investments. For example:

- Rather than dealing with DR as something that is forced upon them, utilities should lead the way to VPPs as part of their portfolio, where and when appropriate.
- Rather than resist the move to distributed resources, utilities need to incorporate all forms of DER into a new portfolio-based business model, one in which the customer (or prosumer) is treated as an ally rather than as a competitor.

Regulators also must get on board to change the regulatory paradigm that today rewards "only iron in the ground" with the ability to get compensated for some R&D, EE, and increased use of DER. For example:

- Regulators should explore retail choice for their state, if they don't already have it, and should allow other players to deliver services to the customers.
- Regulators should also figure out how to incorporate DER into their incumbent utility systems and identify mechanisms with which all can be compensated fairly and equitably.

The advent of new technology requires new thinking as an electric utility considers its future. Developments in communication technologies, monitoring and sensing equipment, DG, ES, and dynamic customer engagement programs require utilities to reconsider the nature of grid operations and resource planning.

We believe that the right course of action for a utility is to embark on a Strategic Business Plan and Technology Roadmap development program via Scenario Planning which looks at all threats, technical and business, and to develop a viable plan which takes the inputs from various stakeholders, including those from potential competitors, customers, and regulators.

Scenario Planning is a powerful tool for utilities to learn to use. They are particularly useful in developing strategies to navigate the kinds of extreme events we have recently seen in the world economy. Scenario Planning enables the utility strategists to steer a course between the false comfort of a typical business planning forecast and the confused paralysis that often ensues in challenging times when many external markets start shifting at once. When well executed, scenarios boast a range of advantages—but if not developed and used correctly, they can also be traps for the unwary.

Given the rapid changes taking place in the energy industry today, a utility without a vision, a sound strategic plan, and a technology roadmap for transformation as to what possible alternative futures it might face is at a competitive disadvantage, even if vertically integrated and/or fully regulated.

When we titled the first edition *The Advanced Smart Grid* in 2009, we purposefully subtitled the book "Edge Power Driving Sustainability." Our previous work at Austin Energy (Chapter 4) and in the Pecan Street Project (Chapter 5) gave us unique insights on the impacts of a shift to DER and on the challenges of adapting to climate change. We recognized that three key drivers would cause inevitable change in the power industry (Chapter 1), whether it would be evolutionary or revolutionary remained to be seen from our perspective five years ago. First, the march of technology innovation, driven by Moore's Law (digitization) and Metcalfe's Law (networking) would enable grid modernization. Second, advancing technologies and scale economies would also enable innovative nonutility companies to develop nongrid energy solutions, which we termed edge power, and which we now call DER. Third, traditional fossil fuel generation would face challenges over the medium to long term as technology-based energy gained a foothold and policy support. After cataloging early smart grid activity in the first edition (Chapter 6), we concluded the following:

> We see an industry evolving towards an integration and network focus (Chapter 2), and we even see hints of recognition concerning smart convergence (Chapter 3), creating innovative industries with new bundles that run along a continuum of increasing integration and market value. The combination of DR, storage, and solar PV (DR-PV) is seen in VPPs and microgrids, emerging components of an advanced smart grid that will be described in more detail in Chapter 7. A VPP system consists of interconnected rooftop solar PV systems (or any other form of DG), interconnected HEMS and DR, ES, and management/control systems. With a VPP, the utility can substitute an integrated bundle of DER to create a new capacity resource that becomes a dispatchable asset equivalent to a traditional peaking power plant (e.g., natural gas peaking power plant). Finally, a microgrid is comprised of those elements of a VPP, with more robust DER systems added to enable increased reliability and true energy independence. As will be discussed in Chapter 7, a microgrid can intentionally island itself from the grid when appropriate, for economic reasons (e.g., a request for curtailment by the grid operator) or for reliability reasons (e.g., imminent outage).
>
> The U.S. market for HEMS, solar PV, and, on a more general basis, DER and smart grid is rapidly evolving, with each of these individual elements in early growth stages. While significant growth is anticipated in all the sectors associated with renewable energy and smart grid, sustained growth remains heavily dependent on governmental policies (e.g., investment tax credits, renewable energy standards, and so forth).
>
> While the ES market may be the smallest market now and in the near term based on technology challenges and relative immaturity, it will become the most disruptive when technology advances finally bring costs down to an affordable level for mass adoption in the second half of this decade. Still, each of these individual elements will require integration into a new, transformed energy ecosystem, as described in Chapters 1–5.
>
> A final conclusion concerns state regulatory processes, which present a challenge to regulators and utilities alike when it comes to smart grid planning and execution. Traditional cost recovery and rate of return regulation provide control parameters on utilities and attempts to balance the interests of stakeholders in an electricity ecosystem made up of regulators, utilities, consumers, and technology providers. Current rules and processes focus mostly on volumetric kilowatt-hours and the supply side, and have not yet fully incorporated demand side resources and the consumer. However, as shown in this chapter, utilities need new motivations and guidance in this highly dynamic economic climate, characterized on one hand by rapid and disruptive technological changes that create new value opportunities, and on the other hand by the need to incorporate consumers into a historically supply-oriented paradigm.

Perhaps many of these conclusions seem more obvious by now, even unremarkable. However, in 2015, as we write this, what is still sorely lacking in industry planning and execution, even in analyst commentary, is a shift of perspective away from gradual changes that allow evolutionary adaptation, toward a more radical proposition. In fact, the three drivers above—technologies for smart grids, technologies for nonutility solutions, and climate change—represent a fundamental restructuring of the energy value proposition, not unlike what we saw from the rapid rise of the Internet 20 years ago and the emerging disruption of smart phones and platform economics less than 10 years ago. When the rules change rapidly, incumbents face unique challenges; many if not most in the power industry are still playing by the old rules.

Early efforts to build smart grids over the past five years or so confirm the first driver—the time has come to build smart grids. As those projects proceed, however, it is rapidly becoming apparent that transformation is not exclusively about technology anymore. Our analysis in Chapters 2 and 3, still highly relevant regarding technology change, underscores the need to adapt in other areas: business model adaptation, organizational adaptation, consumer engagement, and regulatory leadership. As technology projects progress, these four elements pose growing risks of stalling or limiting the potential value of the project. We must note that almost all the reporting in the smart grid space remains rooted in what we might call the old paradigm, focused on the grid and the utility industry. Financial reporting especially remains principally focused on electricity as a commodity, where least cost is the principal attribute (e.g., LCOE). However, in the emerging new paradigm, which we'll explore further in Chapter 7, services and value based on innovation reign supreme. All the benefits of the grid today will remain necessary—reliable and affordable commodity kilowatt-hours to feed plug power, intensive voltages to feed industry, and electric utilities to operate and maintain the grid—but these imperatives must now make room for the higher purposes of service, value and innovation, each key elements of the twenty-first century economy.

In conclusion, let's boldly examine the key points of business transformation using the tools and vocabulary that have become more common after 20 years of Internet and mobile telephony innovation. First, we must conclude that an intensely dynamic business environment demands flexibility and adaptation as core competencies. Understanding ideas like the incumbent's curse (see "Inertia, paralysis, inhibitions, and rigidities" above) and big bang disruption (sudden radical changes based on platform economics) [167] should help us to realistically assess the prospects for the electric utility industry, and to determine the following:

Conclusion #1. Disruption from technology will accelerate and grow larger, driving a greater sense of urgency to adapt. Rapid technology changes will continue to outpace and challenge fundamental aspects of the electric utility industry, posing mounting and incessant operational and business challenges. Time is not on the side of the electric utility industry, but the corresponding sense of urgency necessary to adapt to change is still widely lacking.

In keeping with that theme, we turn to our second trend area, policy. Regulators and policymakers in key cities, states, and regions valiantly craft new models, but their efforts are too often inadequate to meet the challenges of the day, much less the challenges of the next several years and decades. We need more experimentation, bolder policies, and an opening of the utility industry to outside stakeholders more experienced with the pace of technology innovation and competition. Rather than top-down direction and tight control, policy should focus on supporting emergent creativity, openness, standards, innovation, and prudent risk-taking, all fundamental core competencies of successful companies outside the utility industry. As new technologies find niches to support the needs of building owners and managers and consumers, a traditional focus limited to power generation and distribution infrastructure inevitably overlooks new perspectives and alternative funding opportunities. Thus, we arrive at the following:

Conclusion #2. Policy must shift to incorporate DER opportunities and impacts and demand-side perspectives. We must be more creative in our approaches to energy, moving away from traditional mechanisms that have served us well, but now limit our ability to adapt and innovate. We need a broader approach to meet our challenges. The circle of policymaking should expand beyond the traditional electricity industry to include emerging stakeholders with new perspectives on energy, notably commercial businesses; building owners and management firms; large power consumers (industry); local governments; retail services companies including telecom, Internet, and entertainment; and consumers themselves. Just as we owe an undying debt of gratitude to the engineers and business innovators who created, built, and operated the complex and wonderful twentieth-century power industry, we owe the future an equal debt: We must seek out and foster the creativity of new innovative thinkers to map our trajectory into the twenty-first century world of power, which will be equally complex and wonderful, but more plentiful, dynamic, and democratic.

As we accelerate our adaption to keep up with technology change and open up our discussion to a wider audience to encourage creativity and foster innovation, we must also take a long hard look at business models. Every type of business model should be on the table. The idea of a DSP offered up in the NY PSC REV discussion is well thought out and highly relevant as a workable framework for a transition, loaded with potential and notable for leaning on the market to determine the details. We need more of this type of thinking. DSP, after all, is but one perspective of a future business model for utilities. For example, over the past few years distribution system operators (DSOs) blossomed in Europe to address proliferating DER. The DSO is similar but distinct from a DSP model. Moreover, others believe that the future lies in utilities becoming service companies with product/service portfolios based on new technologies. Rather than a top-down directive on a singular business model for the future, the path with the best potential may be for regulators to provide the environment for innovation to blossom, enabling utilities and nonutilities to operate more in a market economy that allows experimentation and risk taking as New York regulators have done, creating a plethora of new business models along the way. So we summarize as follows.

Conclusion #3. Utilities must move from top-down monopoly business models to emerging market business models that leverage risk, innovation, and adaptability in order to meet customer demand for services based on increasing value. The imperative of new business models depends on market creativity, which requires market conditions that support innovation and risk taking. A managed transition is needed to shift from top-down frameworks that focus on maintaining control, limiting risk, investing in assets, and preserving incumbency to bottom-up guidance that fosters greater creativity and innovation and accepts greater risk. It is more important to propel the vital power industry into the future with clear guidelines than it is to get everything right the first time. In a highly dynamic environment, accurate predictions are nearly impossible, so mistakes are inevitable. Edison showed us that progress requires failure in order to learn, leading ultimately to success. Incumbents require a safe harbor not only to continue in the vital role of maintaining the grid and providing services essential to society and the economy, but also to experiment and take risks, to reinvent the industry they know so well. Electric utilities must be free to become more like their new competitors if they are to survive and find more sustainable roles in the new energy economy.

In conclusion, we should all take heart that we have seen tremendous progress over the past five years, but we also must embrace the changes we know are coming rapidly toward us—and the hard work ahead. Our forefathers invented this industry, but they had the relative luxury of time on their side, taking decades where today we have years or months. We no longer can choose to avoid risk through long deliberate evolution; rather, we must open up to new ideas and new stakeholders, challenge our old ways of thinking, and put new ideas to the test in the marketplace of innovation.

At the start of 2015, SG1 continues, if slowly, with grid modernization efforts that must speed up if they are to keep up. SG2, the marriage of old *utility* with new *edge power,* is fast upon us and must be embraced. Meanwhile, SG3, the fulfillment of the potential for positive energy buildings, energy roaming, and beyond, welcomes us to the door and beckons us to come in.

Endnotes

[1]   Apologies to our colleagues at any other utilities inadvertently left off this list.
[2]   http://www.cleanedge.com/reports/reports-trends2003.php.
[3]   http://www.oe.energy.gov/DocumentsandMedia/EDPR2004_session3_von_dollen.pdf.
[4]   Ibid.
[5]   Ibid.
[6]   Admittedly, since most utilities building smart grids are not in market competition, characterizing the trend as a race with a starting gun is really just a metaphor.
[7]   http://www.recovery.gov/Pages/default.aspx.
[8]   http://www.scientificamerican.com/.
[9]   http://www.newsweek.com/blogs/the-gaggle/2010/10/18/locke-goverment-moving-fast on-smart-grid-technology.html.
[10] http://www.nytimes.com/cwire/2010/05/25/25climatewire.
[11] http://www.nist.gov/index.html.
[12] http://www.nist.gov/smartgrid/.
[13] http://www.nist.gov/smartgrid/upload/nistir-7628_total.pdf.
[14] http://www.gridwiseac.org/.
[15] http://www.whitehouse.gov/administration/eop/ostp
[16] http://www.whitehouse.gov/sites/default/files/microsites/ostp/pcast-energy-tech-report.pdf.
[17] http://www.oe.energy.gov/DocumentsandMedia/EISA_Title_xIII_Smart_Grid.pdf.
[18] http://www.energy.gov/organization/labs-techcenters.htm

[19]   http://www.netl.doe.gov/about/index.html

[20]   http://www.nrel.gov/.

[21]   http://www.lanl.gov/.

[22]   http://www.sandia.gov/.

[23]   http://www.lbl.gov/.

[24]   http://www.pnl.gov/.

[25]   http://www.sgiclearinghouse.org.

[26]   http://www.renewableenergyworld.com/rea/news/article/2010/05/where-the-wind
        blows-and-sun-shines?cmpid=WindNL-Wednesday-May19-2010.

[27]   http://www.fcc.gov/.

[28]   http://www.ferc.gov/about/ferc-does.asp.

[29]   http://www.broadband.gov/download-plan/.

[30]   http://www.ferc.gov/legal/staff-reports/06-17-10-demand-response.pdf.

[31]   http://www.naruc.org/about.cfm.

[32]   http://www.naruc.org/News/default.cfm?pr=77&pdf.

[33]   http://www.nist.gov/smartgrid/.

[34]   http://collaborate.nist.gov/twiki-sggrid/pub/SmartGrid/.

[35]   http://www.usnap.org/.

[36]   http://grouper.ieee.org/groups/scc21/2030/2030_index.html.

[37]   http://osgug.ucaiug.org/default.aspx.

[38]   http://www.openadr.org/.

[39]   http://www.zigbee.org/.

[40]   http://www.homeplug.org/home/.

[41]   http://www.wi-fi .org/.

[42]   http://www.nasuca.org/archive/index.php.

[43]   http://newsroom.accenture.com/news/consumers+reject+lower+energy+use+as+the+answe
        r+to+reducing+reliance+on+fossil+fuels+and+energy+imports.htm.

[44]   http://smartgridcc.org/members.

[45]   http://www.nasuca.org/archive/index.php.

[46]   http://my.epri.com/portal/server.pt?.

[47]   http://mydocs.epri.com/docs/Portfolio/PDF/2010_P174.pdf.

[48]   http://www.gridwise.org/. Other national organizations that promote the clean tech industry
        from a broader perspective include the American Council on Renewable Energy(ACORE)
        (http://www.acore.org/front) and the American Energy Innovation Council(AEIC) (http://
        www.americanenergyinnovation.org/).

[49]   http://www.gridwiseac.org/.

[50]   http://www.demandresponsesmartgrid.org/about-ads.

[51]   http://www.demandresponsetownmeeting.com/.

[52]   http://www.pikeresearch.com/research/demand-response.

[53]   http://www.seia.org/.

[54]   http://www.pluginamerica.org/.

[55]   http://electrificationcoalition.org/index.php.

[56]   http://www.itsa.org.

[57]   http://www.electricitystorage.org/ESA/home/.

[58]   http://www.eei.org/Pages/default.aspx.

[59]   http://www.eei.org/meetings/Meeting%20Documents/2010-10-20-StrategicIssues-
        Smart%20Grid%20Scenario%20Project-%20Jahn%20-%20EEI.pdf.

[60]   http://www.publicpower.org/.

[61]   http://www.nreca.org/.

[62]   http://www.utc.org/utc/about-utc.

[63]   http://www.nrtc.coop/pub/us/.

[64]   http://www.edf.org/home.cfm.

[65]   http://www.smartgridnews.com/.

[66]   http://www.intelligentutility.com/.

[67]   http://www.energybiz.com/.

[68]   http://www.renewablesbiz.com/.

[69]   http://www.fortnightly.com/.

[70]   http://www.elp.com/index.html.

[71]   http://www.elp.com/index.html.

[72]   http://www.ieee-pes.org/publications/ieee-power-energy-magazine.

[73]   http://www.renewgridmag.com/page.php?8.

[74]   http://www.renewableenergyworld.com/assets/newsletter/.

[75]   http://www.fierceenergy.com/.

[76]   http://www.fiercesmartgrid.com/.

[77]   http://www.greentechmedia.com/.

[78]   http://www.greenbiz.com/topic/smart-grid.

[79]   http://www.utilitydive.com/.

[80]   http://expous.all.biz/gridweek-expo13074.

[81]   http://www.distributech.com/index.html.

[82]   Itron Looks Beyond the Smart Meter http://www.greentechmedia.com/articles/read/itron-looks-to-more-than-smart-meters.

[83]   Echelon Sells Smart Grid Business to Austria's S&T http://www.greentechmedia.com/articles/read/echelon-sells-smart-grid-business-to-austrias-st.

[84]   Silver Spring Plans Layoffs Amidst a Smart Meter Slowdown http://www.greentechmedia.com/articles/read/silver-spring-plans-layoffs-amidst-a-smart-meter-slowdown?utm_source=Daily&utm_medium=Headline&utm_campaign=GTMDaily.

[85]   Salt River Project Molds Smart Inverters to Ease Solar Transition http://www.utilitydive.com/news/salt-river-project-mulls-smart-inverters-to-ease-solar-transition/302324/.

[86]   SolarCity and Tesla: A Utility's Worst Nightmare? http://www.greentechmedia.com/articles/read/SolarCitys-Networked-Grid-Ready-Energy-Storage-Fleet.

[87]   Why electric cars are selling in California – they're free http://www.thestreet.com/story/12262655/3/why-electric-cars-are-selling-in-california-theyre-free.html.

[88]   It's electric: Experts split on the future of EVs http://www.cnbc.com/id/101918055.

[89]   Tesla bets its patents on a vision of a battery-powered future http://ww.autonews.com/article/20140614/OEM06/306169940/tesla-bets-its-patents-on-a-vision-of-a-battery-powered-future.

[90]   Nevada wins the Tesla battery factory giga-race: Massive incentive package raises questions about corporate welfare. http://www.hcn.org/articles/nevada-wins-the-tesla-battery-factory-giga-race.

[91]   The Local and Global Impact of Tesla's Giga Factory http://www.greentechmedia.com/articles/read/The-Local-and-Global-Impact-of-Teslas-Giga-Factory.

[92]   New eBay Data Center Runs Almost Entirely on Bloom Fuel Cells http://www.forbes.com/sites/heatherclancy/2013/10/02/new-ebay-data-center-runs-almost-entirely-on-bloom-fuel-cells/.

[93]   Bloom Adds Energy Giant Exelon as Capital Source to Finance More Fuel Cells http://www.greentechmedia.com/articles/read/Bloom-Adds-Energy-Giant-Exelon-as-Capital-Source-to-Finance-More-Fuel-Cells.

[94]   Grid Concerns Driving Industrial CHP http://www.fierceenergy.com/story/grid-concerns-driving-industrial-chp/2014-02-04#ixzz2tHADVgZt.

[95]   A Snapshot of New York's Distributed Energy Landscape. http://readme.readmedia.com/
         NYISO-Issues-Distributed-Energy-Resources-Outlook-for-New-York-State/9737185.

[96]   Nanogrids: The Ultimate Solution for Creating Energy-Aware Buildings? http://www.
         greentechmedia.com/articles/read/nanogrids-the-ultimate-solution-for-creating-energy-
         aware-buildings.

[97]   Demand Response Management Systems (DRMS) http://w3.usa.siemens.com/smartgrid/
         us/en/demand-response/demand-response-management-system/pages/demand_response_
         management_system1019-6647.aspx.

[98]   A Glimpse into the DERMS Vendor Ecosystem http://www.greentechmedia.com/articles/
         read/a-glimpse-into-the-derms-vendor-ecosystem-and-solutions.

[99]   http://www.ledsmagazine.com/articles/2011/08/mckinsey-releases-lighting-market-report.
         html.

[100]   LED Lighting: Market Shares, Strategies, and Forecasts, Worldwide, 2013 to 2019. http://
         www.researchmoz.us/led-lighting-market-shares-strategies-and-forecasts-worldwide-
         2013-to-2019-report.html.

[101]   GE Opens a Pricing War Over the Connected LED Light Bulb http://www.greentechmedia.
         com/articles/read/ge-opens-pricing-war-on-connected-led-bulb.

[102]   New Cost Analysis Shows Unsubsidized Renewables Increasingly Rival Fossil Fuels.
         http://www.greentechmedia.com/articles/read/5-more-charts-that-prove-wind-and-so-
         lar-just-keep-getting-cheaper?utm_source=Daily&utm_medium=Headline&utm_
         campaign=GTMDaily.

[103]   "IRP Order".. Integrated Resource Planning (Docket No. 2012-0036, Order No. 32052
         http://lintvkhon.files.wordpress.com/2014/04/decision-and-order-no.-32052.pdf The Ha-
         waii PUC rejected Hawaiian Electric Companies' Integrated Resource Plan submission
         and, in lieu of an approved plan, commenced other initiatives to enable resource plan-
         ning and proffered a white paper entitled, "Commission's Inclinations on the Future of
         Hawaii's Electric Utilities." The white paper outlines the vision, strategies and regulatory
         policy changes required to align Hawaiian Electric Companies' business model with cus-
         tomers' changing expectations and state energy policy; and provides specific guidance for
         future energy planning and project review, including strategic direction for future capital
         investments.

[104]   "RSWG Order" Reliability Standards Working Group, Docket No. 2011-0206, Order
         No. 32053, http://lintvkhon.files.wordpress.com/2014/04/order-no.-32053.pdf The Ha-
         waii PUC makes various rulings regarding the final work product of the working group,
         and provides the PUC's observations and perspectives regarding integrating utility-scale
         and distributed renewable energy resources in a reliable and economic manner; and directs
         Hawaiian Electric Companies, and in some cases, the Kauai Island Utility Cooperative, to
         take timely actions to lower energy costs, improve system reliability and address emerging
         challenges to integrate additional renewable energy.

[105]   "Demand Response Policy Statement" Policy Statement and Order Regarding Demand
         Response Programs, Docket No. 2007-0341, Order No. 32054 http://lintvkhon.files.word-
         press.com/2014/04/order-no.-32054.pdf This order provides specific guidance concerning
         the objectives and goals for demand response programs, and requires Hawaiian Electric
         Companies to develop a fully integrated demand response portfolio that will enhance sys-
         tem operations and reduce costs to customers.

[106]   "MECO Order" Maui Electric Company (MECO) 2012 Test Year Rate Case, Docket No.
         2011-0092, Order No. 32055,http://lintvkhon.files.wordpress.com/2014/04/decision-and-
         order-no.-32055.pdf This order accepted the PUC consultant's report reviewing MECO's
         System Improvement and Curtailment Reduction (SICR) plan and directed MECO to file a
         Power Supply Improvement Plan to address the SICR plan's shortcomings.

[107]   Reforming the Energy Vision (REV) http://www3.dps.ny.gov/W/PSCWeb.nsf/ArticlesByTit
         le/26BE8A93967E604785257CC40066B91A?OpenDocument.

[108]   How a Top Liberal State is Creating an Electricity Market that Con-
         servatives   Should   Love   http://www.greentechmedia.com/articles/read/

could-new-yorks-grid-reformation-plan-be-a-rallying-force-for-conservatives?utm_source=Daily&utm_medium=Headline&utm_campaign=GTMDaily.

[109]   Report to the Department of Public Utilities from the Steering Committee. http://magrid.raabassociates.org/Articles/MA%20Grid%20Mod%20Working%20Group%20Report%2007-02-2013.pdf.

[110]   E.ON may be forced to close gas power plant http://www.powerengineeringint.com/articles/2013/01/EON-may-be-forced-to-close-gas-power-plant.html.

[111]   RWE sheds old business model, embraces transition http://www.energypost.eu/exclusive-rwe-sheds-old-business-model-embraces-energy-transition/.

[112]   Coal Returns to German Utilities Replacing Lost Nuclear http://www.bloomberg.com/news/2014-04-14/coal-rises-vampire-like-as-german-utilities-seek-survival.html.

[113]   Obama's Clean Power Plan Will Get Scrubbed Down Before It Is Enacted http://www.forbes.com/sites/kensilverstein/2014/09/16/obamas-clean-power-plan-will-get-scrubbed-down-before-it-is-enacted/.

[114]   Will President Obama's Clean Power Plan Fly? http://spark.fortnightly.com/fortnightly/will-president-obamas-clean-power-plan-fly?page=0%2C0.

[115]   Japan's Grid Focus: Dealing with Nuclear Implosion http://www.energybiz.com/magazine/article/303951/japan-s-grid-focus.

[116]   Transforming the Power Grid: Lessons from Germany and Japan http://www.greentechmedia.com/articles/read/Transforming-the-Power-Grid-Lessons-from-Germany-and-Japan.

[117]   Transforming the Electricity Portfolio http://www.brookings.edu/~/media/Research/Files/Reports/2014/09/transforming%20electricity%20portfolio%20renewable%20energy/Transforming%20Electricity%20Portfolio%20web.pdf.

[118]   Most new residential solar PV projects in California program are not owned by homeowners http://www.eia.gov/todayinenergy/detail.cfm?id=12991.

[119]   New China Energy Policy Focuses on Distributed Solar and Innovative Financial Tools http://www.greentechmedia.com/articles/read/New-China-Energy-Policy-Focuses-on-Distributed-Solar-and-New-Financing-Tool.

[120]   Solar Capacity in Central America Set to Soar http://www.pv-magazine.com/news/details/beitrag/ihs--solar-capacity-in-central-america-set-to-soar_100016384/.

[121]   California New Law on Demand Response http://www.fierceenergy.com/story/ca-clean-energy-bill-unanimously-passes-senate-assembly/2014-08-28.

[122]   Is an Energy Storage Tsunami about to Hit California? http://www.greentechmedia.com/articles/read/is-an-energy-storage-tsunami-about-to-wash-over-california.

[123]   HECO Aims to Triple Rooftop Solar to Get to 65% RE by 2030 http://www.utilitydive.com/news/heco-aims-to-triple-rooftop-solar-get-to-65-renewables-by-2030/302846/.

[124]   Hawaii Wants 200 MW of Energy Storage for Solar, Wind Grid Challenges http://www.greentechmedia.com/articles/read/hawaii-wants-200mw-of-energy-storage-for-solar-wind-grid-challenges.

[125]   Large Scale Energy Storage to Reduce Load in NYC http://www.greentechmedia.com/articles/read/Grid-Scale-Energy-Storage-to-Reduce-Load-in-New-York-City.

[126]   Duke Energy to Invest $500 M in Solar http://www.utilitydive.com/news/duke-energy-to-invest-500m-in-solar/309312/.

[127]   Austin Texas Passes a New Law Making Solar a Default Generation Source http://www.greentechmedia.com/articles/read/austin-energy-solar.

[128]   Price tag for Austin's renewable energy goal: $1 billion over 10 years http://www.mystatesman.com/news/news/local/price-tag-for-austins-renewable-energy-goal-1-bill/nhTfW/?icmp=statesman_internallink_invitationbox_apr2013_statesmanstubtomystatesmanpremium#eae09d9c.3589834.735501.

[129]   In Electricity War, it's Boulder vs. Xcel Energy. http://www.govexec.com/state-local/2014/07/boulder-colorado-xcel-energy-utility-municipalization/88612/.

[130]   In Vermont, a milestone in green-energy efforts http://www.washingtonpost.com/national/health-science/in-vermont-a-milestone-in-green-energy-efforts/2014/09/14/9fc6e2c6-3c28-11e4-a430-b82a3e67b762_story.html.

[131]   How Sacramento's Public Utility is Getting Into the Residential Solar Business: The New Utility Business Model May Simply Be Answering Customer's Questions http://www.utilitydive.com/news/how-sacramentos-public-utility-is-getting-in-the-residential-solar-busines/301840/.

[132]   RMI Report August 2014 Rate Design for the Distribution Edge http://www.rmi.org/elab_rate_design.

[133]   ALEC: Reforming Net Metering: Providing a Bright and Equitable Future http://www.alec.org/publications/net-metering-reform/.

[134]   SEPA: Net Metering: Many Directions for a National Debate http://www.solarelec-tricpower.org/utility-solar-blog/2014/june/net-metering-many-directions-for-national-de-bate.aspx.

[135]   Is a Value of Solar Tariff (VOST) Really Better Than Net Metering? http://cleantechnica.com/2014/03/29/value-solar-tariff-vost-formula-benefits-solar/.

[136]   It's Time for Grid Planners to Put Distributed Resources On Par With Transmission http://www.greentechmedia.com/articles/read/its-time-for-grid-planners-to-put-distributed-resources-on-par-with-transmi.

[137]   California's Fowl Problem: 10 Ways to Address the Renewable Duck Curve http://www.greentechmedia.com/articles/read/10-ways-to-solve-the-renewable-duck-curve.

[138]   Hawaii's Solar Grid Landscape and the Nessie Curve http://www.greentechmedia.com/articles/read/hawaiis-solar-grid-landscape-and-the-nessie-curve.

[139]   Hurricane Sandy One Year Later: Rebuilding Stronger, More Resilient Communities http://energy.gov/articles/hurricane-sandy-one-year-later-rebuilding-stronger-more-resilient-communities.

[140]   How Microgrids Helped Weather Hurricane Sandy http://www.greentechmedia.com/articles/read/how-microgrids-helped-weather-hurricane-sandy.

[141]   NYC's Hudson Yards Project: Sustainability Exemplified https://nyintl.net/story/hudson_yards_sustainable_building_milestone.

[142]   Grid 2020: Towards a Policy of Renewable and Distributed Energy Resources. http://resnick.caltech.edu/docs/R_Grid.pdf.

[143]   Disruptive Challenges: Financial Implications and Strategic Responses to a Changing Retail Electric Business http://www.eei.org/ourissues/finance/documents/disruptivechallenges.pdf.

[144]   Creative Destruction and the Natural Monopoly 'Death Spiral': Can Electricity Distribution Utilities Survive the Incumbent's Curse? http://papers.ssrn.com/sol3/papers.cfm?abstract_id=2418076.

[145]   Efficiency and Solar Create 'Difficult Road for Global Utility Sector in 2014 http://www.greentechmedia.com/articles/read/ubs-analysis-efficiency-and-solar-create-difficult-road-for-utilities.

[146]   Solar Is On The Way To Dominating The Electricity Market, And The World Has Elon Musk To Thank http://www.businessinsider.com/goldman-on-solar-and-elon-musk-2014-3.

[147]   Citigroup Says the 'Age of Renewables' Has Begun http://www.energypost.eu/age-renewables-begun-solar-power-continues-shoot-cost-curve/.

[148]   The Economics of Grid Defection http://www.rmi.org/electricity_grid_defection.

[149]   What Wal-Mart, Microsoft, and Johnson & Johnson Want From Their Utilities http://www.utilitydive.com/news/what-wal-mart-microsoft-and-johnson-johnson-want-from-their-utilities/307457/.

[150]   Microsoft Announces New Wind Project to Power Data Center http://www.forbes.com/sites/kellyclay/2014/07/16/microsoft-announces-new-wind-project-to-power-data-center/.

[151]   Uploaded to Daimler: Car To Go Acquires Ride Sharing Smart Phone Apps Ride Scout and MyTaxi http://www.pfhub.com/daimler-acquires-ride-sharing-smartphone-apps-ridescout-and-mytaxi-1126/.

[152] http://www.greentechmedia.com/articles/read/ECOtality-Bankruptcy-Blink-EV-Charging-Network-Changes-Hands-But-Not-Bad-R.

[153] http://www.greentechmedia.com/articles/read/electric-vehicle-networks-open-or-proprietary.

[154] NRG and Green Mountain Power Plan to Join Forces http://www.bloomberg.com/news/2014-09-02/nrg-green-mountain-plan-vermont-renewable-power-network.html.

[155] ComEd leading community-based microgrid http://www.businesswire.com/news/home/20140916006179/en/ComEd-Awarded-Grant-Department-Energy-Microgrid-Controller.

[156] Mini Solar Becoming Big Business http://www.fierceenergy.com/story/mini-solar-becoming-big-business/2014-09-12.

[157] Energy and Carbon Benefits of Pico-Powered Lighting http://lightingafrica.org/wp-content/uploads/2014/08/Issue-4_EnergyandCarbonBenefits_Final.pdf.

[158] Institute for Local Self-Reliance (ILSR) http://www.ilsr.org/

[159] Communities for Advanced Distributed Energy Resources (CADER) http://www.cader.org/.

[160] A How To Guide for Transactive Energy http://www.greentechmedia.com/articles/read/a-how-to-guide-for-transactive-energy.

[161] PowerShift Atlantic http://www.greentechmedia.com/articles/read/powershift-atlantic-the-virtual-power-plant-of-the-future.

[162] Sharp Introduces SmartStorage Energy Management Solution: Immediately Available in California, with Expanded Availability Planned for 2014. http://www.sharpusa.com/AboutSharp/NewsAndEvents/PressReleases/2014/July/2014_07_29_SmartStorageEnergyMgmnt.aspx.

[163] Property tycoon reveals $20-billion solar-led portfolio http://www.theglobeandmail.com/report-on-business/industry-news/energy-and-resources/property-tycoon-reveals-20-billion-solar-led-portfolio/article20860227/#dashboard/follows/.

[164] Siemens CEO Discusses How Knowledge Becomes Money http://www.forbes.com/sites/robertreiss/2014/07/28/lessons-on-innovation-from-visionary-ceos/.

[165] http://www.512cmg.com/.

[166] Disruption Becomes Evolution: Creating the Value-Based Utility, CMG, 2014 http://www.512cmg.com/wp-content/uploads/2014/07/CMG-Creating-the-Valued-Based-Utility-30-June-2014.pdf.

[167] Big Bang Disruption, Harvard Business Review, March 2013 http://hbr.org/2013/03/big-bang-disruption/.

CHAPTER 7

Fast-Forward to SG3

The comprehensive review of smart grid events in Chapter 6 confirms and demonstrates many of the advanced smart grid concepts and arguments found in Chapters 1–5. With a keen focus on edge power, the insightful observer witnessed an expanded vista of emerging advanced smart grid concepts revealing themselves from 2012 to 2015. Three principal trends noted in Chapter 6 paint the picture: (1) Smart grid technology expanded well beyond AMI to include rapidly accelerating DER adoption; (2) regulators, policymakers, and utilities struggled to craft new policy and responses to DER; and (3) business models emerged as a fresh topic in this discussion, but regulated utilities and nonregulated power companies lagged private service companies in the level of innovation and activity, with entrepreneurial technology companies acting far more aggressively in crafting new business approaches based on new DER technology capabilities.

Introduction

In this chapter, as we wind up this thread of discovery that we call the advanced smart grid, we briefly review the key concepts presented in this book, we reveal the complexities inherent in the advanced smart grid, we show how these concepts and complexities are addressed in novel planning methodologies and architectures, and, finally, we take a longer look out into the next decade and beyond, to see where the advanced smart grid journey will lead us as SG3 begins to emerge.

Our review includes a discussion of the role a *SGAF* can play as a how-to book on building the advanced smart grid. Also, we describe the visionary new *SGOE*, which will be needed to operate the advanced smart grid using real-time updates from all managed devices correlated with all utility and end-user systems and to coordinate all the emerging new edge devices in the IoT.

Thus, we'll explore in detail the methods of planning, designing, and operating advanced smart grids. Once a decision is made to begin the advanced smart grid journey, the utility is on a path of constant, incremental innovation. The advanced smart grid changes the rules of the game in a fundamental way. The traditional utility business model is punctuated by episodic events and projects that occur between periodic, if sporadic, rate cases. In contrast, the future will feature more rapid change that will necessitate more frequent rate cases or new regulatory oversight tools. We may witness alternatives to rates as the principal means to finance structural change. For example, self-financing programs may tap personal, corporate, and real estate wealth to structure payments according to a new financial rhythm. On the far end, we may expect jurisdictions to begin to unbundle and deregulate different functions within the vertically integrated utility to adapt to the emerging dynamic market.

To conclude our journey, we look at the transition to SG3, where a vision of a clean, linked future involves such innovations as transactive energy, energy roaming, and integration of ES. When pervasive IP networks and computing and energy are commingled with abundant information and new renewable energy and DER technology, new forms of energy trading will become as common in the future as accessing content over the Internet is today. The future holds the potential for plentiful clean energy managed over sustainable robust networks.

We have termed this ambitious future SG3, which we believe represents a golden age of abundance, where we manage what we have with greater respect for limits and boundaries, but we also get greater use and enjoyment from what we have, thanks to sustainable networks that eliminate or minimize waste and encourage easy, even effortless transactions. SG3 will come from applying the concepts and principles elaborated in the previous chapters. As these ideas become incorporated into standards and templates, utilities will gain a practical approach to change that balances current short-term needs for reliability and continuity with longer term needs of rationality and sustainability.

Deploying an advanced smart grid, in our vision, offers a way to become far more efficient using natural resources. For example, if we have the potential today to implement an advanced smart grid and, through grid optimization, reduce the total system energy losses, from 20% down to 10%, then that delta of 10% represents conserved natural resources that could be saved for use by future generations. Innovation is at the heart of the exciting message of the advanced smart grid.

With regard to grid modernization in SG1, our view is that smart grid architecture and integrated network design, rather than the application, should be paramount in smart grid planning. And regarding DER in SG2, we believe that policymakers and the utility industry should acknowledge clear trends and plan for a future that seamlessly blends DER with centralized assets, with utilities using the grid for economic as well as physical balancing of resources and loads. Finally, the concepts outlined in SG3 are waiting to be picked up and demonstrated by entrepreneurs in utilities and in private companies as technologies and consumer markets mature.

Looking Back

The Inevitable Emergence of the Smart Grid

In our review of the principal concepts in the advanced smart grid vision in Chapter 1, we used real-world scenarios to demonstrate the foundational arguments for an advanced smart grid and to showcase to the world a way for adopting these principles. We began by showing how much of the complexity in early smart grid projects derives from a decision to begin the project at Layer 7 in the OSI stack, the application layer, which requires significant system integration and limits future potential. We offered an alternative perspective to begin with a deliberate smart grid architecture design. The advanced smart grid will emerge because electricity is the fundamental energy source for the twenty-first century and because it is the nature of technology to empower consumers. Technological advances make our lives more convenient and place individuals squarely at the center of decisions that affect their lives. Most technological change, including DR and, in a broader sense, DER integration, will recede into the background over time, as technology is configured according to the "set it and forget it" method.

The Rationale for an Advanced Smart Grid

Three principles highlighted in Chapter 2—security, standardization, and integration—drive the creation of an advanced smart grid. As digitization becomes normative, the challenges associated with implementing security and the vulnerability of the grid will require an advanced smart grid to ensure implementation of sufficient security measures. The standardization of digital devices and networks enables low-cost and rapid adoption and supports advanced smart grid development. Finally, the proliferation of digital solutions and their gradual integration in networked architectures will bring about the emergence of the advanced smart grid as the logical and most efficient architecture.

The traditional functional silos of electric utilities—generation, transmission, distribution, and retail services—in the twentieth century, will gradually dissolve as a utility reorganizes its operations as an integrated energy ecosystem. Moving from vertical to horizontal organization in this way will enable utilities to leverage the capabilities of digital control systems over networks and operate more efficiently. Vendors, too, will need to adjust to this new way of doing business. Historically organized to sell products and solutions into those silos, vendors will begin to shift their focus to a fundamental redesign of the grid to harness the digital revolution and engage new thinking about architecture and network design based on lessons learned from the Internet. This transformation to an *energy Internet* capable of routing power and information in much the way that the Internet routes bits and bytes today will take years to implement, informed as it is by a long-term vision that accommodates a dramatically different set of needs. However, the need to bring consumers into the picture through DR and to integrate DER will beg the question: *How will the system be kept in balance with millions of devices integrated into the grid from all points?*

Integration will be a huge challenge. After all, we can't continue to depend on adding ever more generation-based solutions to provide the balancing services needed to accommodate intermittency. This brings to mind still more questions:

- How will tens of thousands of HEMSs cycling hundreds of thousands of appliances on and off be enabled in a distribution grid that is currently blind to activities that lie beyond the distribution substation?
- How will the distribution grid add multiple EV charging stations that appear in a neighborhood over the course of a few months, when the transformer located at the end of a distribution feeder was designed decades ago to manage a static load limit based on the number of houses it served?

- How will the system accommodate ES units when technology matures over this decade and makes ES an economical solution?
- As grid parity approaches, how will bedroom communities with multiple rooftop solar installations feed their excess power back onto the grid while residents are off at work during the day, when the grid is designed for one-way power flow?
- How will grid managers address grid stability when large amounts of intermittent energy from wind and solar farms are added?

Each of these questions highlights the need for a more advanced smart grid to address the new problems of increasing complexity. Moreover, these questions carry the discussion far beyond the relatively less complex prospect of transitioning from analog meters to smart meters to provide interval data to support TOU rates or the more complex project to add sensors and controls within the distribution grid to improve visibility of grid conditions during outages. As necessary as those steps may be, they offer only a partial solution to the bigger problem that those questions point to: *The grid is becoming far more complex and demanding, and technology solutions only buy time—transformation is needed to meet the long-term demands of complexity.*

Smart Convergence

The story of progress in the modern world has been a story of collapsing innovation cycle times, as technologies enable information and innovation sharing and bring innovators closer together, actually accelerating the pace of change. The utility infrastructures that support modern lives were mostly built 100 years ago give or take a decade or two, and now they are on the path to digitization. As similar upgrades make infrastructures smarter, ideas are adapted from one industry to the next, and patterns begin to emerge, leading to what we call smart convergence. In the meantime, as smart convergence itself matures, barriers between industries collapse still further. The advanced smart grid will revolutionize electricity grids just as the Internet transformed the supporting telecom network.

Case Study: Smart Grid Enterprise Architecture, Integrated IP Network(s), and SOA

In Chapter 4, we used the case study method to detail the processes, organizational issues, and lessons learned from building the first smart grid in Austin, Texas, from 2003 to 2010. A critical success factor in that pioneer endeavor was recognizing the importance of a smart grid enterprise architecture design process that started with customer engagement and service goals. Focusing on the customer strategy and associated needs allowed the enterprise to appropriately document the necessary processes, which, in turn, could then be supported by the appropriate underlying infrastructure, data, and application strategies and architecture design. The key lesson learned from that experience was that any other approach would result in some combination of higher costs, greater complexity and risk, and/or diminished ability to optimize the customer interface. This lesson learned came from trial and error, but also out of the internal struggle between IT and OT and the need to manage to a strict budget.

Envisioning and Designing the Energy Internet

The answers to such integration questions are found in the vision of an energy Internet, described in detail in Chapter 5. The energy Internet is designed to be resilient and robust, sufficient to support both the current and future demands of integration. The challenge facing grid owners and managers today demands nothing less than a return to the drawing board to design an infrastructure capable of meeting our twenty-first century needs. With a smart grid architecture design in place, effective project planning and project management are needed to transition to the future state while maintaining system reliability.

Chapter 5 reviewed the exhaustive community brainstorming that took place in the Pecan Street Project in 2009 to move beyond the current utility paradigm to a new concept for an energy production and delivery system. The resulting Pecan Street Architectural Framework and the other tools, processes, and concepts are instructive for other communities and utilities as they plan their own futures in a highly dynamic environment.

The final lesson learned introduced in Chapter 5 is the need to engage customers. Making the community more aware of electricity production and consumption issues and of potential impacts on economic and environmental outcomes will strengthen the odds for consumer acceptance of energy technology innovations and the adoption of advanced smart grid programs, associated pricing signals, and dynamic rates. Consumer engagement will require a period of education and internalization of new energy use habits as a prerequisite. The shift to an advanced smart grid introduces dramatically more information into the ecosystem, raising the profile of data privacy issues, data storage and management, and the value and importance of effective communication and operations between the utility and its customers.

Today's Smart Grid

Recent events described in Chapter 6 reveal an energy world that is rapidly expanding to include a multitude of new, nonutility stakeholders, driven by steady advances in modular energy technologies collectively referred to as DER, which we called Edge Power in the subtitle to this book. We discerned three principal trends that we used to highlight these changes. First, DER technologies are advancing faster than expected, upsetting the status quo. Such disruptions challenge grid operators to adapt to new stresses to maintain reliability, just as they challenge utilities to craft new ways to replace revenue lost to DER solutions and nonutility service providers. Second, policymakers as well as pioneer utilities struggle to engage DER with grid modernization plans under current paradigms. The best new policies create altogether new paradigms that more closely align with DER. Third, business model adaptation, or lack thereof, is under discussion inside and outside the utility industry. Nonutility companies especially are adopting new business models to compete better with utilities, challenging the incumbents to respond.

Advanced Smart Grid Complexities

The next two sections describe a variety of new complexities and associated capabilities that will be needed to manage the advanced smart grid. The first section focuses on grid operations and the second on market operations.

Grid Operations

Resource Islanding

Resource islanding, or "islanding" for short, is a term used to describe the voluntary or involuntary off-grid functioning of a premise, community, or local area that has the capability to provide power for itself. Learning to master the islanding of power and load is a necessary step to fully understand how to manage dynamic DER on the grid.

The potential disruption to grid operations and synchronization caused by islanding poses a vexing challenge to grid operators, which have traditionally viewed islanding as more of a bug than a feature. Few entities have the capability for voluntary islanding today, and notwithstanding the disruptive elements of islanding, mastering this capability would be an attractive option were it to be made available. Let's talk first about *voluntary* islanding and the challenge of synchronization. Consider a university that has developed a grid capability and then decides to disconnect voluntarily from the grid during a peak period and operate on its own microgrid. How is the now islanded university going to reconnect onto the grid after the peak, much like stepping off a moving train and stepping back on? The synchronization wave inside the current requires a smooth transition to synchronize the waves of the two systems. The physics of electricity make timing vital.

The sine wave in our 60-Hz grid completes a full cycle from the bottom of its curve and back in about 17 milliseconds (i.e., $1/60 = 16.66667$ milliseconds). The synchronization of two separate electrical systems is a complex task to reengage electricity flow at the sine wave level. The larger system determines the reconnection protocol; the university microgrid must conform to the utility grid status at the point of connection. To accomplish reconnection, the two entities first synchronize their concept of time, then the grid operator defines the point in time when the switch is closed and both grids reconnect. The grid operator must reengage the system deliberately, either automatically or manually, at the point when the two grids are synchronous, to keep the system in balance. If connecting an islanded microgrid to a utility grid, the microgrid must have an automatic generation controller like a power plant has and conform to rules for synchronization and reconnection.

The connection and disconnection of wind farms involves a similar challenge. Over time, similar issues of engaging and disengaging to the grid will arise from the greater market penetration of DG. Integrating significant amounts of DG will dramatically multiply the potential impact of these synchronization scenarios, as innumerable small resources are added at different spots on the grid with increasing frequency.

Islanding presents many questions, including the following:

- If a feeder is set up to handle 1,000 kW, can the operator add more than that with an islanding event? What considerations are needed for grid operations planning?

Answering these questions will enable more microgrids to be added, as well as VPPs, renewable energy facilities, DG, prosumer engagement and control, and regional grid balancing during volatile events that disrupt grid harmony. The bottom line is that gaining islanding capabilities makes system management more flexible: Load can be reduced to maintain grid harmony and grid operations on the regional grid, and normal operations and full power conditions for those within the islanded region. During an historic cold snap through much of the United States in February 2011, ERCOT had to order rolling blackouts in Texas to keep the grid in balance. Power plant operations at more than 80 power plants had been disrupted by the extreme, sustained cold weather, reducing voltage levels on portions of the ERCOT grid. Furthermore, as more power plants went offline during several days of intense cold weather, healthy portions of the grid that were still producing power were ordered by ERCOT to reduce their production to preserve regional grid stability.

Almost a year and half later, Hurricane Sandy hit the populated East Coast in late October 2012, bringing with it both high winds that knocked down outdoor grid facilities and a storm surge that swamped underground utility facilities. With damages totaling over $68B, this storm would prove to be the second most costly in U.S. history, surpassed only by Hurricane Katrina in 2005 ($108B). As the winds died down and recovery efforts got under way in November, the press highlighted microgrids on the East Coast that continued to provide service during widespread blackouts.

Had islanding been available to grid operators in Texas, Austin Energy's operations would have avoided the rolling blackouts it endured to maintain ERCOT's grid stability, and had it been available in the northeast, billions in economic losses may have been avoided. Fast-forward a decade and such blackouts will be avoided when ISOs have islanding protocols and when utilities have programs that encourage individual and collective islanding. Islanding, an interesting and economic solution for enhancing grid stability, awaits an advanced smart grid to realize its potential.

Dynamic Modulation

When we talk about grids that operate at 120V or 240V or 480V to their edges, at either 50-Hz or 60-Hz frequency (the United States runs at 60 Hz), we refer to midpoints in a range and to grid management operations that keep system voltage within that range. Dynamic modulation is a microstrategy for fine-tuning the grid in real time to manage grid frequency and voltage levels with sensors for more efficient resource and load operations and enhanced reliability. Smart edge devices and algorithms enable such tightening, tuning, and toning of the grid. Dynamic modulation will provide grid operators a new tool to replace manual processes to manage the range in voltage, ensuring smoother operations.

Current manual processes will be overwhelmed as complexity increases, making automation a necessity. The manual multitasking required by current grid operations are akin to managing a fleet of plate spinners out on a football field [1]. The spinning plate management task would move to 10 then 100 football fields and would also expand to include innumerable small backyards and parking lots. Further, the plate spinners would not be trained, professional plate spinners but amateurs intent on meeting their own needs, expressing their creativity according to their own whims and desires. If that sounds like chaos, it is. And it certainly could turn out that way without dynamic modulation and other tools and well-defined rules of the road to automate grid management and ensure that myriad and dispatchable DER devices are orchestrated to balance the grid and optimize new resource capacity. The advanced smart grid will be needed to bring order to this impending chaos.

Predictive Volt/VAR Control (PVVC)

PVVC is to predict and maintain acceptable voltage at all points along the distribution feeder under all loading conditions. The power electronics associated with PVVC must account for the constant imbalance and rebalancing of the grid. Power factor is the measure of volt/VAR specific to a location where these events impact grid conditions. Advanced power factor is the capability to adjust the ratio between inductive and conductive loads.

Starting with a perfect power factor of 1.0, the grid becomes less and less efficient as power factor declines, requiring more voltage to bring the power factor levels up. Thus, low power factor from uncorrected imbalances at the load level makes grid operations more expensive. PVVC—the ability to improve power factor—will require less voltage. Before that can happen, however, power factor must be measured. Homes in the future will gain the ability to export both volts and VARs as they gradually shift from a load-only state to a combination of loads and resources.

PVVC offers grid managers a new, real-time capability to accommodate increasingly complex grid conditions to analyze and control load changes in real-time. Such control will focus on capacitor banks, voltage regulators, and load tap changers (LTCs) to manage the volt/VAR relationship. PVVC will be used to flatten the load profile that goes through the feeder—to push the volts and VARs towards equivalence—a goal of a power factor of 1.0—thereby minimizing line losses and optimizing distribution feeder functionality. More detailed information about system voltage and VAR levels will more precisely define the current grid state, allowing more defined local directives as a new tool for power factor correction.

PVVC brings multiple benefits, including reduction of line losses, reduced generation and carbon footprint, an ideal power factor 24/7/365, and a flattened feeder voltage profile. PVVC will act in multiple ways: capacitors injecting VARs, generation-adding voltage, and DR-reducing loads to avoid expensive peaks.

Predictive FDIR

Like PVVC, predictive FDIR—also called FLIR and FLISR—provides a new capability to address grid conditions, this time by using prediction to improve recovery when things go wrong. FDIR is comprised of three active steps: (1) fault detection (through monitoring), (2) fault isolation (through control), and (3) restoration to normal grid conditions (through management). One day, digital devices attached to edge devices will accomplish these tasks automatically, according to preprogrammed algorithms. An advanced smart grid will provide a self-healing component that is lacking today with switches, transformers, and capacitor banks enabled with FDIR capability. Today when a device fails, the failure may go undiscovered until another event occurs and the discovery process detects, isolates, and restores the fault. Predictive FDIR is paramount to manage disruptive events including involuntary islanding, with rapid corrective actions by the utility that prevent damage to expensive equipment and associated costs.

These new capabilities are needed to address the increasing complexity of the grid and maintain grid reliability and operations. Adding predictive FDIR will reduce outage duration, frequency, and restoration times, typically measured in such indices as SAIDI, SAIFI, and CAIDI. Predictive capability will enable tighter system management and lengthen system asset life, and a logged system of events will provide an accounting record heretofore unavailable, which will improve management practices over time.

Demand Action

By now, DR is a well-understood term in the electric utility lexicon, a utility mechanism to control load-using such devices as HEMS and smart thermostats. In contrast, demand action may be seen as a utility mechanism for the dynamic dispatch of DER resources, an edge power supply corollary to DR. As new localized resources proliferate and the grid optimization engine is deployed (see below), grid operators may reach out and use demand action to increase the supply of power to benefit efforts to balance the grid at specific sites. With abundant and widely distributed DER, a utility could automatically stimulate energy production in a targeted fashion (at specific sites) using a price signal and offer to buy.

IoT

IoT is the logical progression of the pervasiveness of the Internet reaching traditional utility infrastructure devices not network-enabled until now and not controlled by computer systems until now. Some companies call it the industrial Internet (i.e., GE) and others call it the Internet of everything (i.e., Cisco). IoT is the interconnection of uniquely identifiable embedded computing devices that can be managed by leveraging the Internet infrastructure. Utilities already have their own private networks and are also buying many services from public telecom carriers, including Internet access. So imagine now utilities leveraging computing infrastructure like mobile phones (e.g., iPhone or Samsung Galaxy) and allowing employees to get notification and even control embedded devices managed by their own networks and now interconnected to the public carrier networks. Typically, IoT is expected to offer advanced connectivity of devices, systems, and services that goes beyond M2M communications and covers a variety of protocols, domains, and applications. The interconnection of these embedded devices is expected to usher in automation in nearly all sectors of the economy and society, while also furthering advanced applications like smart grids. A current example widely used by utilities includes the remote monitoring of smart thermostats and home appliances with Internet access.

Market Operations

Abundant Information

The condition of abundant information describes a new condition where a plethora of data-gathering devices heretofore unavailable are deployed and busy producing mountains of data that must be stored and protected. Analysis of abundant data with data analytics tools to produce valuable and abundant information will guide improved grid management on both the supply and demand side. Abundant information also implies the open sharing of information previously isolated in silos, away from computers, people, and devices that could have used that information.

Utilities will need to share much more information with their partners and producers out on the grid's edge. Where utilities once operated as isolated power experts managing a well-defined grid under controlled conditions, with degrees of limited information, they will shift in the future to operating as collaborative power experts managing a less defined grid, but with far more information and automation. Abundant information will not only involve more data available on the load curve, but also DER producers gaining access to the data needed to better understand their own generation equipment—to do predictive maintenance, for example. Utilities will be able to use this new, abundant information to provide new energy services to create new types of value. A leaky feeder, for instance, may produce altered volt/VAR conditions that require adjustments to get the most from the active and reactive power that the DER produces. The cooperation and integration challenge facing utilities today concerns finding ways to open up to selective sharing of information when they have traditionally protected information for privacy and security reasons.

Prosumer Control

Prosumer control will provide the consumer who both consumes and produces energy control over new DR and DG responsibilities. The consumer maturity model (see Figure 7.1) demonstrates the path that a consumer must go through to move from a state of passivity and lack of awareness, up through successive states of evolution, to transform into a responsive, committed partner of the utility in helping to manage a more rational approach to energy production, distribution, and consumption. At the most mature level, the consumer has transformed into a united, committed, networked prosumer, engaged in building supportive communities and innovating with new energy technologies.

As this consumer maturity model shows, early utility programs track maturity levels if they are to communicate effectively with consumers diverging into different market segments with different levels of maturity. Passive, unaware consumers with information to increase their awareness of impending changes and causal relationships require orientation to become aware of new possibilities. Later, newly aware consumers will gradually become more active with experimental pilots hosted by the utility. As they mature, consumers will shift from monitoring energy usage to actively managing their consumption, curtailing load on demand and hopefully providing load data to the utility. With aggregated behavior and more experience, the customer becomes a utility partner, now responding to utility

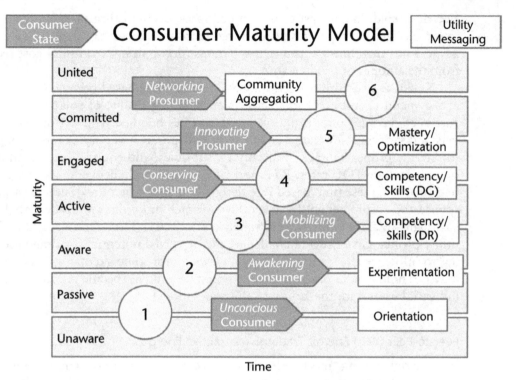

Figure 7.1   Customer maturity model.

requests to shave peak and participating in utility rebate programs. With the integration of consumer-owned DER systems, the consumer becomes a prosumer, enjoying utility payments for edge-based power and ancillary services. Finally, innovating and networking regionally will bring the new prosumer to his or her most advanced state, in which he or she is partnering with peers and other stakeholders to redefine the energy ecosystem.

In the transition to an advanced smart grid and mature consumers, *consumer engagement* represents a tremendous challenge for utilities, which have traditionally seen marketing as relatively one-dimensional, principally outbound information distribution via bill stuffers (analog), Web sites, and social networking applications (digital). A growing awareness among utilities will lead to change, but such change will require extensive resources and considerable time. Households and businesses comprised of passive and unaware consumers will not mature overnight into united and committed prosumers. Nevertheless, as they mature over time, their transformation will have an immense impact on the future prospects of an advanced smart grid.

Dynamic Pricing

New dynamic pricing will be reflected in new rate structures that recover fixed costs more equitably, using new, unbundled rates that provide greater distinction regarding cost recovery and greater transparency on separate services, providing the necessary first step for a utility to recover fixed costs regardless of the amount of electricity sold. In contrast to unbundled rates, decoupled rates are an adjustment to rate design that separates utility revenue from the amount of energy sold, using a rate-adjustment mechanism that separates (decouples) fixed utility cost recovery from the amount of product sold.

New rate structure options may include: (1) time-based pricing; (2) fixed rates and demand charges; (3) solar customer class (e.g., value of solar tariff); (4) EV customer class; (5) pilot programs; and (6) rates bundled with various DER products and services. New rate designs must incorporate a variety of rate-design tools to find the proper balance of incentives, subsidies, and equity. Time-based pricing options include TOU rates, RTP, and CPP. A central challenge to successful adoption of any of these time-based rate designs will be to elicit the desired customer response and ensure benefit for both the customer and utility; it is a delicate balance to design a rate that allows a customer to see electric bill savings yet still maintains utility profits. Given such a challenge, the perceived potential for time-based pricing to influence energy use should be viewed with some skepticism. Assumptions may not hold true in practice or may hold true only in specific markets (e.g., large industrial users with the ability to shift operations).

Peer-to-Peer (P2P) Energy Trading: Transactive Energy

P2P energy trading, more recently referred to as transactive energy, envisions a world of edge-based energy transactions, where producers of energy on the edge use their grid connections to "ship" power to another party, where the utility provides merely transport services (this economic transaction is similar to "retail wheeling" in the 1990s). Please see the following SG3 section below for a more complete discussion of P2P energy trading/transactive energy.

Revolutionary Smart Grid Tools: SGAF and SGOE

Moving on from this discussion on smart grid complexities and new capabilities, let us shift our focus now to two new smart grid tools for the design and operation of advanced smart grids. First, the SGAF provides grid designers with a handbook of best practices, rules, and methodologies on how to build an advanced smart grid. Next, the SGOE is a revolutionary new tool to operate the advanced smart grid, bringing optimization to both the distribution grid and to the DER elements connecting directly to the grid and indirectly to buildings.

SGAF: A How-To Guide for the Advanced Smart Grid

An SGAF is a set of standards, best practices, rules, and methodologies to build a smart grid architecture—answering the how question. As described in Chapter 4 in the discussion on smart grid architecture design, architecture is comprised of the following components: infrastructure (networking, security, computers, and data storage systems); data; applications; and processes. Where the design is the written plan and blueprint to direct activities, the architecture framework (standards, best practices, rules, and methodologies) is the set of instructions on how to build the smart grid. The framework is the cookbook by which one builds a smart grid; the architecture itself is the artifact that describes the smart grid.

Building a smart grid is a new science that requires the application of knowledge and experience. NIST, IBM, and Microsoft have provided architecture frameworks on how to build a smart grid that remain the dominant models used to design smart grids in this emerging industry. The set of methodologies in SGAFs encompass a wide variety of disciplines including standards; systems to use (databases, operating systems, programming language, hardware, computers); requirements on storage; and disaster recovery. Beyond their instructional value, SGAFs are valuable for recording the successes and lessons learned from past efforts, thereby documenting the growing wisdom in the industry.

The use of SGAFs for planning represents a vital tool as we enter a new, experimental space. Smart grid is a new science, and like every science, there is a significant amount of art involved. *Where does a smart grid designer go for advice on how to build a smart grid?* Expertise is found in pockets throughout the industry, and SGAFs are emerging over the past several years, with NIST, IBM, and Microsoft showing leadership. These SGAFs establish the standards, methodologies, and best practices to bring together power, telecommunications, and software and hardware systems to address the needs of an energy ecosystem made up of central generation, transmission and distribution lines, and meters and customer systems and beyond the meter, DER: DR, EVs and charging infrastructure, DG, ES, and smart appliances.

An SGAF would help a smart grid designer assemble, on average, over 100 systems in a rational and effective way to create, distribute, and consume energy more reliably and affordably. Philosophical designs from traditional information technology evolution are transforming the old way of designing the grid, helping smart grid designers to leverage the best lessons learned from past successful transformations in the telecom and computer industries.

A key challenge of the emerging SGAF is that, by default, each electric utility has its own architecture in place. A designer must forge a path to transition from the old architecture to the new architecture in real time. As we have said, the way to build a smart grid will ultimately involve both science and art—techniques and adaptations will work together to craft successful projects. Given the slow adoption of SGAF, few can claim true knowledge of the optimal methodology, much less claim the experience of having gone through this process and come out the other end. So we are all learning from each other, and the evolution of the SGAF is showing us the way. The following sections describe the domains and systems associated with an SGAF.

Central Generation

The management of power plants has always been highly automated, as have been the processes and functions around the power plant's output. Unique systems in this domain include automatic generation controllers (AGCs), time series databases, energy hedging systems, wholesale scheduling and settlement systems, generation management systems, distributed control systems, emissions management systems, and laboratory management systems.

Transmission Lines and Systems

Transmission system components have become highly automated as well. The most familiar elements are supervisory control and data acquisition (commonly, SCADA) and EMS, often referred to together, as EMS/SCADA systems. Other prominent systems in this area include load planning systems, asset management systems, and, more recently, emerging synchrophasor systems. This new domain promises a new level of analysis for grid operators, akin to having an X-ray of a point in time that allows detailed data correlation, which will help define a new level of decision-making and management of the transmission system. Phasor management units (PMUs) and the software algorithms to correlate the data output will comprise a new synchrophasor system. This new tool is increasingly being integrated into transmission management protocols, promising new powerful levels of management and control at both the transmission and distribution levels.

Distribution Lines and Systems

This domain has historically been the focus of the evolution of the smart grid. Distribution infrastructure lacks the full automation that the prior domains (generation and transmission) have enjoyed over the past half century. As the grid has evolved and the variability and complexity of load has increased, the need to automate distribution lines and associated infrastructure has come to define what we talk about when we discuss smart grid. Key systems in this domain include DMSs, OMSs, GISs, asset management systems (AMSs), vegetation management systems, load planning systems, workforce and workflow management systems (WMSs), and mobile workforce systems, as well as the OT areas of SA and DFA.

Meters and Customers

The genesis of smart metering started with automated meters and drive-by AMR and a supporting business case to replace pedestrian manual meter readers with automatic readers in vans. While early AMR technology has been eclipsed by two-way fixed-wireless AMI, deployment of AMI remains unfinished business—in the United States, AMI penetration stands at 50M out of 140M and elsewhere in the world, AMI is still a significant opportunity as a growth market. The ability to provide Web-delivered services and information came with the advent of the Internet, augmenting the value proposition of AMI. The smart meter remains an evolving trend, as does the way customers receive information on their energy consumption habits. Some unique systems in this domain include MDM systems, customer portals, Web services, call center interactive voice response (IVR), online billing, online self-help, conservation management systems, DR systems, marketing program and customer loyalty systems, and with the smart phone, the emergence of mobile energy apps to engage the customer in greater visibility and control of their personal energy.

Beyond the Meter: Home and Office Systems

This domain remains the Wild West of the utility space. Solutions beyond the meter may be provided by the utility or by competitive third parties. While the utility demarcation point has traditionally terminated at the meter, a growing attitude among utilities sees "beyond-the-meter" no longer a hard and fast rule, but a relic of days past. The smart thermostat has opened the door for greater utility involvement, and now pool pumps, water heaters, lighting, and other edge devices are considered fair game for new utility services. Pioneering utilities in "beyond the meter" services in the NA utility market include EPB (Chattanooga, TN); City of Provo, UT; and New Brunswick Power (NBP). Utilities in Australia and South Asia are moving aggressively into multiservice provisioning, combining electricity with an array of telecom, entertainment, and security services.

The emergence of an ever more mature private energy management industry offering BEMSs, HEMSs, and many more services is now being characterized with the terms "connected home" and "smart building." New and existing companies that will provide these services may include security companies, gaming and computing companies, telecom providers, and cable operators, to name just a few. Key systems that will accelerate dramatically in this space are customer portals, Web services, call center IVR services, billing services, DRMSs, and marketing program and customer loyalty systems. A new set of systems is emerging, including EV charging systems, smart inverters (PV), DER management systems (DRMS, DEMS, DERMS, and ILM), and ES management systems.

As with any new industry, the value of standards is paramount. Evolving standards to guide the creation of these SGAFs range from interoperability to security. The central catalyst for the development of standards comes from NIST, which is empowered by FERC to develop a collaboration between a large number of organizations, including IEEE, IEC, ANSI, SAE, OASIS, and UCA, to name a few. Modbus, the original SCADA and PLC communications standard, has been joined by such key standards as IEC 61970 and 61968 (application-energy management system interfaces), IEEE C37.118 (phasor measurement unit communication), DNP3 (substation feeder automation), IEEE 1547 (inverter standard for physical and electric interconnection with DG, EV and ES), IEC 6870 (intercontrol center communications), Open HAN (home area networks), ZigBee Smart Energy Profile 1 & 2 (home energy management communication and data), BACnet (building automation), ANSI C12 (metering), OpenGIS and OAGIS (GIS), and, finally, IEC 62351 (information security for power system control). Of note, IEC 61850 specifies substation automation out to the meter, providing guidance on the link between edge devices and the substation. NIST remains the best resource for a complete list of evolving standards in the smart grid industry.

In the emerging area of smart grid-smart building convergence, the OpenADR (IEC/PAS 62746-10-1) standard for price response and direct load control is increasingly evident for integrating with building automation systems (BASs), gateway devices, and more recently, cloud-based services that provide remote control of commercial, industrial, and residential controllers/devices. ILM and advanced DRMS and DERMS technologies are designed to accommodate any devices in compliance with OpenADR 2.0. IEC 60870-5-104 is used for generic load control, as

MultiSpeak is used for load control through AMI head-ends making architectures more extensible. The worldwide KNX standard is used by more than 250 manufacturers of products that optimize lighting, shading, heating, and cooling in rooms and buildings. Intelligent building networks use EN 50090 and ISO/IEC 14543.

SGOE: From Static to Dynamic Grid Operations

The advanced smart grid will require a new way not only to model the behavior of electricity flow on the grid, but also a way to operate the advanced smart grid in real time. Currently, the electricity system is highly stable and predictable. As DER penetration grows, however, distribution grid operations grow more complex (for an example, see Chapter 6's section on the Nessie phenomenon in Hawaii). In essence, grid planners regularly conduct weekly modeling sessions, where they model predicted energy flow on the grid based on historic demand and anticipated weather impacts. With a weekly network electric model (or whatever planning period is needed) in place, reality takes over when operators adjust and manage the grid according to actual load consumption and more immediate data than was available when the plan was devised.

In contrast, the SGOE includes functionality that one would find on a network-modeling tool but focuses on dynamic balancing of volt/VAR levels based on real-time data inputs from a multitude of devices (e.g., smart meters, smart transformers, smart feeders, EVs and charging stations, and smart inverters). However, the SGOE also provides the ability to control the devices and the grid in real time.

How will this differ from what we currently use to manage the grid? First, the SGOE anticipates a much more complex environment, where two-way power flow occurs as the norm rather than the exception. Two key concepts to keep in mind when discussing SGOE are: (1) updating the system to enable real-time grid management (requiring real-time communications capability) and (2) managing in a predictive manner, anticipating failures before they occur (i.e., self-healing processes). As mentioned earlier, we now plan the network electricity flow in advance, and as real-world conditions unfold, we use the new data from one period to update the model for the next period. In this way, grid operators tweak the model as needed with each new planning period, but in an ideal scenario, the SGOE would instead optimize grid operations in real time and drive all aspects of grid management automatically, continuously adjusting the model as data comes in, not waiting until the next week to issue a new plan. The SGOE requires added new equipment to meet this real-time operational vision in a two-way environment (e.g., protective gear in all controlled distribution assets).

The next improvement to the current methodology would be to incorporate what-if scenarios to enable predictive modeling to address gaps. To move into the predictive realm in grid operations, we will need the benefits of *big data* and *data analytics*. Access will be needed to all the relevant infrastructure data, such as device procurement date and installation date to calculate the age of the equipment, service status including repairs and reasons, and redeployment, if any. Such asset management data needs to be incorporated because the quality and capability of the assets determine the self-healing potential of the system.

The main point of the SGOE is to attain a state of self-healing and automated, efficient operations. Furthermore, to realize the potential of SGOE, the interconnected SCADA system needs to provide real-time control capabilities to execute based on defined capabilities and to indicate updates and replacements of relevant assets as needed. We are moving from a current state where maintenance repair crews follow a serial schedule with regular repair cycles, to a dynamic schedule based on actual equipment performance and failure rates, where adjustments and fixes are made just prior to failure or immediately thereafter.

A key benefit from the SGOE is to replace a current inefficient repair and planning system based on stale data with an efficient repair and planning system based on fresh data. The risk of relying on stale data to do repairs, maintenance, and system planning is that the grid becomes overbuilt in places. A "gold-plating" scenario results when stale data leads designers to build the grid for a possible peak event that may never occur. The SGOE provides a much truer picture of grid status that leads to more effective planning and lower capital expenses over time by reducing peak safety building practices via its real-time flattening of the load curve.

Another way to imagine an SGOE is to use the analogy of a utility ERP; what ERP does for organizations today is fairly well understood and documented (i.e., integrated general ledger plus inventory tracking plus asset management and work orders plus purchasing—all these systems contribute and draw from a common database to be able to synch and maintain a common organizational status view). The SGOE needs to lead a similar transition of the complex electric grid to do what ERP did for the enterprise management: create a more integrated operational model and ensure optimal functioning of the advanced smart grid.

To better understand the potential of an SGOE, let's walk through how an SGOE with true control would work as an engine to promote optimized grid operations integrating the utility grid with edge resources. In this scenario, a distribution feeder line is reaching its capacity limit. Currently, there are two principal ways of managing that situation: add more generation to increase the power carrying capacity of the line (increase voltage) or reduce load via some DR or load control event (curtailment). An SGOE would expand those options, for instance, by channeling local energy onto the line from a more immediate source but also by curtailing locally where it was economic and optimal.

This functionality is the focus of new management systems in an edge category that we might label a DER optimization engine (DEROE), which would be a subset of SGOE that includes such systems as HEMS, BEMS, DRMS, DEMS, DERMS, and ILM. These systems coordinate disaggregated resource and curtailment contributions, which enables a complete VPP operational model comprised of not only curtailment, but also edge generation. The full realization of SGOE will be stimulated by the need for situational intelligence and control at the edge.

As the evidence in Chapter 6 shows, the rapid maturation of DER (in particular, solar PV) now has the attention of electric utilities, regulators, and policymakers. Some are doing more than others, but there is plenty of documentation of the disruption that is already occurring. What remains less clear is what to do about this march of alternatives to grid-supplied power, raising questions such as the following:

• How will utilities manage the disruption to system operations? What about the challenge to utility revenues as kilowatt-hour sales dip?

Awareness will be the beginning of control. As the saying goes, "You can't manage what you don't measure." As DEROE capability finds its way into pioneer utilities, we can expect experimentation with changes to the utility business model. It will be vital that utilities engage with DER vendors to achieve technological as well as business alignment. Utilities will need to execute partnering arrangements with these vendors, to cobrand services and bring added value to their customers.

Here's how such technology advances may stimulate new business model approaches. A utility deploys some type of DEROE software, then solicits interaction with DER vendors based on Open ADR compliance. The vendors engage with the utility to gain accreditation and ensure technical alignment. After trials and pilots to ensure functional interoperability, utilities will need to develop comarketing programs. A utility program manager may work with colleagues inside the utility to identify a tranche of DR, an objective to shut down a specific resource, ideally an inefficient resource or a polluting resource. The details of this DR program are communicated to participating vendors, which develop and take offers out to customers gain market acceptance of new value-added services that also target specific load reductions. The customers get value-added services, the vendors get service revenue, and the utility gets enhanced operations and improved profitability by avoiding uneconomic or otherwise costly operations.

With experience, the parties should improve on this basic pattern, getting ever more skilled at reducing peaks. In time, the leveling of load will result in the retirement of inefficient, costly, or dirty energy resources. In time, the leveling of load will reduce system volatility and expenses associated with managing complexity and unpredictability. In time, the utility may be smaller and may have less revenue, but there should be a greater reduction in cost and risk and an increase in profitability to offset the old ways. With the optimization of DER through the DEROE, we can expect a new kind of utility that operates at both ends, using central resources as needed and harnessing distributed resources and dispatchable loads, as needed. The action will take place along the distribution feeder.

The Smart Grid Journey: From 1.0 to 2.0 to 3.0

The transition to an advanced smart grid perspective, as outlined in this book, may be inevitable, but as the comment on VPP and the SGOE above makes clear, the transition is not necessarily going to happen overnight, nor uniformly across regions—nor will it happen without a conscious shift in paradigms. The shift from SG1 to SG2 is occurring as outlined in this book and by our theme of simplifying complexity by shifting focus to a more network-oriented perspective. However, the shift from SG2 to SG3, which we'll discuss in more detail in the following sections, will be based on a full acceptance of technology and a letting go of the forces that hold us back.

Nature's DER

To produce an abundance of energy in highly efficient ways, nature has evolved ingenious and harmonious ways to harness the second law of thermodynamics, most notably such processes as respiration and photosynthesis. Nature does this by going small. The key to this value proposition is the emphasis on optimized, distributed energy production at the microlevel. The system stays in balance by optimal use of energy production byproducts in a closed loop to enable corresponding energy-producing processes.

Animals use mitochondria to produce energy at the cellular level with consumption and respiration (i.e., a human body consumes carbohydrates, proteins and fats, water, and oxygen to manufacture energy at the cellular level, producing waste and carbon dioxide as byproducts). Plants use chlorophyll to produce energy at the cellular level with photosynthesis, by consuming water, minerals, and carbon dioxide and then using sunlight as a catalyst to transform the water and carbon dioxide into glucose and oxygen (e.g., an apple tree).

In this way, nature provides us with a model of an integrated, distributed network to produce edge power efficiently—a tree is a distributed network with some 100,000 micropower plants embedded in its leaves, complete with, among other things, a distribution system and edge intelligence. Similarly, the human body is a distributed network with millions of cells whose mitochondria are micropower plants, managed and controlled by a neural network that automates certain functions as an optimal design. Repeating an analogy from Chapter 1, reflexes cause the hand to rapidly jerk back when touching a hot object, without conscious thought, because the nerve pathway becomes more expedient to route a preprogrammed signal to stimulate a response as a survival mechanism. The advanced smart grid will be designed similarly, with routine and urgent responses preprogrammed in algorithms to intelligent edge devices for optimal performance. For such reasons of efficiency, the combination of automated response and distributed intelligence at the edge is the wave of the present and the future.

How energy production is distributed matters greatly in terms of efficiency. Beyond the focus on sustainability described earlier, nature has crafted a marvelous method of managing risk, widely distributing the means of energy production, so that both production and consumption occurs everywhere (i.e., is widely distributed) and energy resources are located near to where they are consumed. This approach calls to mind the way that we have come to manage risk in the financial markets, via a portfolio approach, where a basket of investments has a lower collective risk than a single large investment. Likewise with energy production, nature makes multiple bets and lets it all shake out, with winners rising to the top like cream in a milk bucket. We draw a lesson from these observations, concluding that in this dynamic environment, we face less risk by moving energy production closer to its point of consumption—and by betting on DG over central generation over time.

SG3 Emerges

By the time SG3 begins to take shape, much will have changed in the utility landscape. Some distribution utilities will have adapted to a proliferation of DER and transformed their grid operations and processes to accommodate hundreds of thousands of new devices, becoming DSOs responsible for market-clearing transactions and grid coordination, or DSPs offering a platform to host market-based service applications and transactions. Those who have not will likely have merged with those who have. The distribution energy revolution truly begins when grid parity arrives and a new energy era is acknowledged and accepted. Utilities that have implemented advanced smart grids will find themselves enjoying the benefits of a flatter load curve and avoided capital expenses for new generation and new distribution system capacity upgrades, which will let them concentrate on new tasks.

As old approaches of peak management via large peakers give way to integrated DER dispatched to manage peak demand at the edge, the utility operating model will have shifted. DER collocated with the load it serves is likely to win any auction for peak consumption, leaving existing peakers to convert to base load service or to be decommissioned. Regulators and policymakers should also discuss this element, which leads to a more complete transformation than some have in mind today. Whether DER is owned by the utility or a third party, the modular nature of DER located next to load becomes the more viable economic dispatch model. Current ROE or ROA targets will become unacceptable in the future, when such alternatives exist.

As the era of SG3 dawns, the advanced smart grid will become not just a way to deliver electricity more efficiently, which will bring tremendous value; it will become an entirely new social and transactional platform. New business models, applications, services, and relationships will emerge to leverage new possibilities and new potential created by the shift to pervasive DER. Just as the Internet slowly worked its magic to transform our lives at home and at work, the advanced smart grid will provide abundant and affordable power where and when it is needed. Regional and local economic and social success will become based on the adoption of advanced smart grids, because the availability of reliable, affordable electricity will become a key differentiator.

In fact, the key challenge for utilities in SG3 will become reinvention to operate in conditions of abundant supply, as the advanced smart grid unleashes the ongoing, incremental addition of DER devices throughout the grid. In some ways, one is reminded of the transformation in the office environment based on the increase in worker productivity after IT became widespread in the enterprise. The current dialogue in terms of supply side and demand side will give way to a discussion of quality of power and QOS, as grid optimization facilitates new electricity market concepts, including *energy roaming* as EVs proliferate, *transactive energy* (i.e., *P2P energy trading also know as transactive energy*), *ES, VPPs,* and *microgrids.*

A Word on Use Cases

The use cases that follow are helpful to show how an energy ecosystem enabled by an advanced smart grid could look in as little as 10 years, when SG3 becomes reality. If the concepts outlined in this book are widely implemented, such fictional visions become the logical projection of potential trends and outcomes, a way to integrate this book's many themes and discussions. Following each use case, we highlight relevant innovations.

Use Case 7.1: Energy Roaming and EVs—Jack and Jill, Up the Hill in Their New EV

In 2020, Jack and Jill love their sophisticated lifestyle in San Diego, one of the most advanced cities in the United States when it comes to electricity. Their host utility, San Diego Gas & Electric (SDG&E), has been at the forefront of research on smart grid for nearly 15 years by 2020, and California has consistently led the nation as one of the most progressive electricity climates.

So, when the price of EVs dropped nearly in half under a new SDG&E program that put the EV batteries under a lease contract with the utility, Jack and Jill decided it was time to take the plunge and bought an EV (the *EV cooperative program* required Jack and Jill to follow certain charging protocols and billed them each month for a *capacity charge*, essentially a lease bill for the batteries, and an *energy charge* for the electricity they used. Every three years, the utility replaced their batteries with fresh, stronger batteries and put their partially depleted EV batteries to work balancing the grid from racks inside distribution substations).

At first, they had been worried about their car running out of juice, known as *range anxiety* in the industry, but their first few months with the new car reassured them. They found that they naturally adjusted their usage patterns to shift their in-town driving to the EV, and for longer distances they used their second car. Instead of *his* car and *her* car, they now had a *town* car and a *country* car. Sure, they had to coordinate a little more, but in this way, they both got to enjoy their new toys as well.

SDG&E installed and maintained the car charger in their garage as part of the capacity charge and offered them three charge levels at different rates. At the fastest charge setting, they could charge their EV in a couple of hours, but they rarely did so because the rates were so high [2, 3]. Just as SDG&E owned the electric meter on the wall of their house, so it owned the charger in their garage [4]. The couple had entered a new rhythm of charging at the lowest rate overnight, which was more than ample to support the 20-mile commute into town, or even a day of running local errands, which could cover as many as 40 miles in a day. While away from the house, they could charge at work, at any of the special charging meters downtown along the street, and in parking garages, at the airport, and increasingly in retail locations like the mall, the zoo, and local restaurants [5]. The EV charging/parking spaces were always in premium spots, up-close—an unexpected perk. Soon after they bought the new car, Jack and Jill had downloaded an app for finding available parking spots for EVs [6]. Charging was as easy as plugging in, not unlike putting a gas pump nozzle into the gas tank. The charging stations had user-friendly interfaces, similar to self-checkout at grocery stores, and an app was also available to automate the identification process of charging [7]. Moreover, thanks to the wide deployment of DSPV systems in San Diego, when they parked in a garage, their power was almost always coming directly from solar PV panels overhead [8].

Another aspect of SDG&E's EV Cooperative Program allowed Jack and Jill to improve the efficiency of the grid by storing and releasing energy very rapidly, without any effect on the battery's state of charge. They programmed their car to a minimum charge level, and during the day, their EV battery could sell energy back onto the grid based on economic signals from the utility [9]. Their power bill provided a detailed list of debits and credits to their utility account, not unlike early cell phone bills [10].

A new program at SDG&E linked with other California utilities, so that Jack and Jill could take their car on the road up to Orange County, timing their charge periods to stay within the 100-mile range per each charge. They didn't do that often, but when they did, the new roaming program let them charge up while in OC and have their home account billed at their home rate [11].

Jack and Jill were happy with their EV. They both liked the savings and the environmentally friendly carbon reduction. Jill liked the leather seats, the incredibly smooth and quiet ride, and the comfort of the cabin; Jack liked the in-dash GPS with RideTracking, Smart Phone dock, and integrated Bluetooth and the downloadable "car tones" that project an exterior sound to warn pedestrians of the EV's presence. Jack liked to step on the accelerator and feel the torque of the powerful electric motor on occasion, near the house so he wouldn't run dead while out and about. The EV, believe it or not, was quicker than the motorcycle he'd had as a teenager. Who knew?

> *EV/ES cooperative program.*   An EV is conventionally viewed as a replacement or substitute for a car with an ICE. However, from a utility perspective, an EV is much more—it is in essence a mobile form of ES. The challenge of ES has vexed utilities for years, as ES technology slowly progresses and prices stubbornly resist dropping the way that PV and LEDs prices have fallen. Long called the game changer of all game changers, ES implementation has eluded utilities, as ES prices have stayed high without tech breakthroughs or the scale production that has nudged prices down for PV and LEDs. Nevertheless, as utilities recognize the value of both EVs and ES, a business model innovation will allow utilities to blend EV and ES programs in something like the cooperative program in the case study above.
>
> Such a program would have two key aspects, one around ES and the other around EV. First, under the ES part of the program, utilities leverage their low cost of capital to buy lithium batteries in bulk, then place them with EVs, recovering a portion of the costs of the batteries through "capacity" charges to EV owners subscribed in the program. At an agreed time (around three years), the utility would replace the batteries with a new, less expensive set and start the depreciation process over again. In this way, the utility will get still viable used batteries at a depreciated cost, which they may locate in stages in racks inside substations or in converted smart transformers to provide ancillary services and to protect utility equipment from the dangers of power backflow. This process is not unlike buying stocks in a portfolio through a monthly payroll deduction, diminishing the risks of market timing, price averaging over time. Similarly, utilities need not time their ES purchase to some time long in the future when ES becomes affordable. They can begin to enjoy the benefits of storage immediately and pay less over time for their ES systems, learning along the way new and different ways to deploy ES and integrate new functionality into their grid operations.
>
> Under the EV part of the program, the utility dramatically drops the costs of EVs for their subscribers by removing the upfront purchase of the battery system from the vehicle price—effectively, unbundling ES from the EV. Dropping the sticker price will accelerate EV adoption rates, bringing load growth into the utility and adding revenues sooner than might otherwise be expected. This type of innovative cooperative program supports a utility strategy to provide more value to its customers and locks them into a firmer longer term relationship, supporting positive brand image.

Finally, responsibility for the life and maintenance of the EV battery system shifts to the utility under the cooperative program. Should the system fail or experience untimely deterioration in performance, the EV owner need simply schedule a replacement at a service provider certified by the utility. Existing automobile dealerships may see this as a valuable new service to offer in coordination with the utility.

V2G.   However, the value of EV batteries is not limited to the acquisition of ES through such a cooperative program. Consider the dramatic value potential that a V2G program could unlock, as described in the scenario below. If just half of the vehicles nationwide were to be shifted to EVs under programs like the cooperative program, bulk energy production could remain the same, and our energy problems would begin to disappear. In 2011, we used the example of an aggregation of parked EVs in Austin, Texas, taking approximately 300,000 EVs with 10 kWh battery each of capacity represents 3,000 MWh, which is roughly equivalent to the current capacity of centralized power production by Austin Energy. Soon, however, EVs will have 100-kWh batteries, reducing the number of EVs needed from 300,000 to just 30,000 to have the same impact on the Austin Energy grid. If each EV is driven on average four hours each day, it is left parked for 20 hours each day. Assuming that the EVs are charging for four hours each day as well, then that leaves 16 hours each day per EV to be part of a distributed, collective ES facility, assuming that each were connected and left actionable as a part of an advanced smart grid. On an average trading day, the going rate for a megawatt-hour in ERCOT is $30 (on a peak day, that number goes as high as $5,000 in 2014 but will move to $9,000 by summer of 2015). In that case, 3,000 MWh of wholesale capacity or 3 GWh for a single hour would be $90,000 (30 × 3,000) on a normal trading day. Meanwhile, in a peak hour at $5,000 the actual cost would be $15 million for 2014 and a shocking $27 million by summer of 2015. Thus, the value of our collective EV storage facility dedicated as a grid asset during its available time, multiplied by 365 days, gives a new annual capacity resource for this utility ranging from $525 million ($90,000*16*365 or $525,6000) for a normal day/hour to $60 million for just four peak hours in 2014 to $108 million for the same four peak hours in 2015. There will be money to be made in a fully distributed energy market where EVs discharge back to the grid any idle capacity.

Energy roaming and EVs.   The explosion of EVs, which will ultimately include new types of vehicles including electric motorcycles and bicycles, will bring the concept of energy roaming into the discussion. Energy roaming calls to mind roaming minutes on a cell phone bill, as used to be the case a few years ago. However, in our case, energy roaming is a positive with payment of a "home" fee regardless of where the energy is consumed, rather than an additional charge on your phone bill.

The possibility of energy roaming could emerge as described in this section as DER becomes more widely adopted. In fact, to put it more strongly, for energy roaming to move from concept to reality, DER must become more widespread. As long as our energy economy is oriented to fossil fuel power plants and low-cost grid power, the payment for energy will remain tied to the fuel.

Energy roaming will require bilateral contracts between utilities and, no doubt, some regulatory changes.

Energy roaming will occur inside the utility service territory as a new service, where the key will be to attach a charging event to a particular account. Austin Energy currently offers a great EV charging program allowing members to charge in any of the 200-plus charging stations for unlimited times for the monthly price of just $5.00. Energy roaming between service territories will be similar to the cellular telecom experience in another way, with a new market created for utility back-office accounting transactions. Energy roaming decouples energy service from the commodity energy sale, making it open to third-party sales and services. Energy roaming will entail Internet access to a platform including new applications and content and new pricing and new service models that depend upon how much energy is consumed (kilowatt-hours), when the energy is consumed (time of day), and where it is consumed (geographic coordinates).

Use Case 7.2: P2P Energy Trading and Consumer Engagement—Home Energy Management Evolves the Customer Role
6–8 a.m.: It's all hustle and bustle as the two kids, Dad, and Mom busily ready themselves for school, work, and a day full of errands and charity work. Thanks to the overnight precooling cycle, the home is cooler than it used to be in the morning, but the family is now used to it—indeed, it's now a great way to start the day. The air conditioner is done for the next several hours—HEMS (see Figure 7.2) has disabled it according to its program—it won't need to start up again until much later

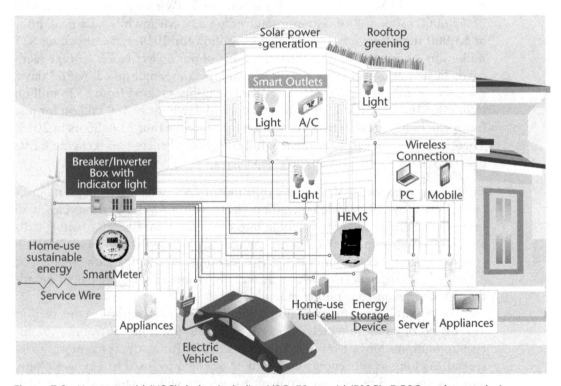

Figure 7.2 Home-to-grid (H2G) design including V2G, ES-to-grid (ES2G), DG2G, and smart devices.

in the day [12]. HEMS plugs and breakers attached to CFL and LED light bulbs in the ceiling and in lamps measure energy usage and ensure optimal usage patterns. Dad scoops up his electronic devices, casually noting that the indicator lights on the HEMS plugs and power strips are glowing red—a good sign, since that means his devices finished charging sometime in the middle of the night and the HEMS devices sensed the change in power draw and shut off power to the transformers, which otherwise would have kept drawing power and wasting electricity—even though the devices were already charged [13]. "So many leaks now plugged by HEMS," Dad thinks and smiles, knowing that those cents have a way of turning into dollars.

Mom is last to leave, casually glancing at the HEMS unit on the refrigerator and seeing that all is well. In a hurry, she locks the door on her way out. She can't help but note that all the lights are off and the house is eerily silent, as if it had gone to sleep until their return (she doesn't realize that Junior left the LED TV powered up, even though the DVR turned off the picture—it's still drawing power). She smiles, knowing that should she realize she forgot something, she can always access the HEMS personalized dashboard on her smart phone or the GUI interface in her EV [14]. "It's taken some getting used to being so in control," she thinks to herself, "but now conserving energy is more a habit than a conscious activity, and besides, so much is programmed and automated by the HEMS, there really isn't much to think about." As if hearing her thoughts, the HEMS plug sensed the lack of use and the time of day and shut off power to the TV screen, one of the home's biggest energy consumers.

8 a.m.–4 p.m.: While the family is away, the HEMS breaker monitoring the circuit on which the family's hot water heater is connected automatically opened the circuit, cutting power to the heater, until the wireless sensor notified the HEMS that it was approaching its minimum temperature and turned the power back on. The HEMS thermostat monitored the air temperature inside their home throughout the day. Thanks to its connection to the HEMS, which is in turn connected to the Internet, the thermostat is aware of events beyond the four walls of the home. For instance, it just learned that a cold front has come through that day, bringing storms and much lower temperatures. Aware of these environmental changes via the Internet and www.weather.com, the HEMS overrode the HEMS thermostat's programmed precooling cycle, which would have started the air conditioner at 2 p.m., similar to its run the night before [15]. Meanwhile, outside, the temperature sensor in the pool failed to register a high enough temperature to make the HEMS breaker engage the pool pump to cycle on to keep the pool temperature at a comfortable, but affordable level—again, because it turned out to be a cool day and the added measure of rain in the pool had kept the water temperature down as well—more energy saved [16].

4–10 p.m.: Mom was first home, with Sis and Junior in their new EV and a trunk full of groceries. As the kids took the groceries inside, Mom plugged the car into the HEMS EV charger in the garage. Mom noted that the indicator light turned to yellow as she plugged the car in; sensing the plug in but aware of the time of day, the HEMS EV charger remained off until lower rates would begin after 10 p.m., when it would power on and begin charging to have the EV ready to go in the morning.

With the kids chopping vegetables, Mom sat down in her kitchen office to review the family's power consumption status on her laptop, hitting the HEMS.com dashboard page. The Web site prompted her for user name and password. She quickly reviewed the family's energy status, glancing over her "pictures"—even though she could tell by the green frame around each that all was well, she still enjoyed spending a moment or two to get a fuller understanding—she called to the kids to come see the jar with black jelly beans spilling over—so much carbon offset in only 6 months, well past their initial goals. It felt good to be a part of healing the planet [17]. She noticed that the dial for the solar PV panels was only slowly turning—cloudy weather will do that—usually it was spinning happily away, registering clean kilowatt-hour production [18]. One last thing, seeing that they were still well ahead of their goal on their projected electric bill, with 10 days left in the month, she tweaked down the total electricity consumption line to provide a little more of a challenge for her family. Now during the last few days of the month, the HEMS warning light would be glowing yellow, urging the family to go the extra mile to squeeze out the last bit of savings. It really was amazing how six short months of data feedback and control devices from HEMS had taught them to change their energy habits and behaviors, while still allowing them all the comfort they wanted. Also, their new detailed consumption data correlated with their total consumption measured by the HEMS—that information came in handy and even had helped them win an argument with their utility two months ago, when their bill was out of synch with their HEMS data—a new experience [19]. "Could conserving energy actually be fun?" she mused…time to get back to making dinner, and get the kids to fold the clothes and empty the dishwasher.

Just then, the phone rang—no doubt it was Dad, calling to say he'd be late from work. After the aforesaid notification of a goal reset, he couldn't help but chide his wife for putting the squeeze on the family. Seems that he'd set his personal alarm to notify him of any negative changes in energy consumption, and his wife's goal tweaking had moved the family into a caution zone and triggered his alarm. Relieved that it was nothing to worry about after checking the HEMS dashboard on his work PC, he humored his wife's desire to get a little more out of their HEMS investment. He reminded her that it had already more than paid for itself just last month, three months ahead of time. She reminded him of his mother, who lived on a fixed income back in New Jersey. Rates were much higher there, and she suggested they could start shipping their savings to her each month as part of the new P2PET program that their utility had recently set up. Impressed by her desire to save, and delighted about this new way to share their savings, he found his new conservation behavior rewarding in more ways than one, not nearly as tedious as he'd expected. "What was so good about wasting energy, anyway?" he thought.

Before he got off the phone, he reminded his wife they were due to look into new curtailment rates that had just become available from their utility. They'd have to spend a little time on the HEMS Web site to see how families in other areas with such rates were dealing with the curtailment option.

Next up, he knew, was the potential to join a group of like-minded HEMS users—they'd been tracking the HEMS blog and were looking forward to their first HEMS neighborhood meeting next Wednesday, when they'd all talk about pooling their accounts to get the same rates that large office buildings did. There would also be a presentation by a group that wanted to form an energy MUD (eMUD) in their area. What was that all about? Were they ready for it? He smiled, knowing they were, and knowing HEMS would help them on their new journey to becoming engaged energy prosumers [20].

10 p.m.–6 a.m.: As the family sleeps, the air conditioner gradually cools the house according to the cooling algorithms that the HEMS learned in coordination with the family (it is summer, and it gets hotter down in the South, even in San Diego). Both refrigerators and the garage freezer are plugged into HEMS plugs, programmed to cycle off and on to provide optimal use of power, but also to skip the chill cycle altogether during a critical peak pricing event [21]. The dishwasher, washer, and dryer all started automatically soon after midnight, on notification from the HEMS that lower off-peak rates had begun (since they got HEMS, the family now loads the machines before bedtime, turns them on, and leaves them, knowing they won't start until told to by the HEMS). Just before the alarm clocks began to go off, the family's pool pump ran one last cycle before shutting down for the day, as instructed by the HEMS breaker attached to its circuit.

> *From net zero energy to positive energy buildings.*   The home described in the case study above may have been renovated or designed from the outset as a net zero energy or positive energy building, two concepts that extend beyond LEED. Over the past decade, our focus on energy efficiency has led to the discussion of houses and buildings that are designed and built to meet new net zero energy building codes. Under this approach to construction, the building, which is equipped with an EMS, anticipates high levels of energy efficiency and is outfitted with connections that enable EV, solar PV, and other forms of DER. The net part of net zero says that over the course of a year, the building will require zero energy, on balance (when each monthly bill is added together to produce an annual demand).
>
> However, a new, more aggressive approach to designing buildings would anticipate the building becoming a power plant itself. As DER prices decline and systems mature, the building would be designed for excess production, a.k.a. a negative home energy rating system (HERS) index, with the size and capacity based on the property's capacity and the owners pocketbook: as much as the rooftop can hold, as much as the property dimensions allow, as much as the market demands, or as much as the owner's budget allows. In other words, the building is by design a PEB. A PEB is a building that is not only a residence or place of work, but also a source of income as a distributed independent power producers (IPPs) (DIPPs). Furthermore, positive energy mortgage loans, where homeowners and businesses apply the surplus they receive from energy sales to accelerate the payoff of mortgages, can be expected as a financial innovation. Imagine a 10-year mortgage that is accelerated based on the income from the PEB. We may go even further to imagine reduced mortgage rates because the added revenue makes payback less risky.
>
> The PEB anticipates a future demand and market for DER in the DSO/DSP. The PEB designs and builds in its production capabilities before the building is even constructed. In SG3, the utility may schedule PEB resources in addition to other DER and centralized resources. Two conclusions from this trend are noteworthy: First, 40 percent of all energy in the United States is consumed by buildings—imagine a world where buildings are built to include power

production—and second, this trend taps into alternative nonutility sources of capital to fund the energy system. Combining energy with real estate will become both a way to increase the value of the real estate and a way to enhance the power system, an example of *smart grid-smart building convergence*. In this win-win scenario, the utility and its ratepayers become less reliant on rates to finance the grid, and carbon mitigation is achieved without bankrupting the economy.

P2PET (a.k.a. transactive energy).   When the combined impact of DG (solar PV), ES, and DR are integrated on the advanced smart grid, and the utility adopts a DSO or DSP role, then P2PET will become a new reality. As the name suggests and this short description makes clear, P2PET involves the electric utility only as a distribution resource. The production on one end and the consumption on the other end are conducted absent utility involvement. Recent discussions about transactive energy follow this concept, envisioning a system where energy is traded at the retail level.

P2PET could also occur over longer distances, from two separate electric utility service territories, much as wholesale transactions are conducted today. If electrons truly are innately fungible, why can't individuals and organizations provide their excess power as credits on an accounting ledger, directing the credit to peers who could enjoy the benefits of "free" energy they consume off the grid? Can electricity one day become a more fungible commodity, where the transaction is decoupled from the electron flow (i.e., transactive energy)? Can electricity be traded the way that airline miles are today, accumulated in an account by a frequent flyer, and then donated to his or her favorite charity at Christmas time? Airline frequent flyer programs have become widely accepted means to build loyalty, and isn't that just what the electric utilities need to be doing?

Trading energy over the Internet the way we currently move content is a matter of accounting, leveraging new grid capabilities brought to the front by the advent of the advanced smart grid. In other words, accounting for inputs and outputs provides a balance with transaction and connection fees. When energy becomes abundant, and systems become more capable, P2PET (as in Use Case 7.2) could become a new energy service offered by electric utilities or new energy service providers. Consider that we are even likely to see microgreen energy producers emerge as a new small business as P2PET takes hold.

Let's talk in more detail about how P2PET or transactive energy could be implemented at utilities. Energy production at the edge will need to be forecast and planned, perhaps by zip code or neighborhood or, more likely, by distribution feeder level. PPAs between the utility and DER owners are a mechanism to accomplish such planning, because they serve to lock in the capacity sale and tie build or buy decisions on the grid to emerging market pricing and retail needs. At the outset, a utility can use a solar rider or similar mechanism to pay a premium for DER, but as DER becomes ever more common, exceeding centralized generation, then the purchase needs to become more dynamic to more closely resemble the current wholesale market behavior that we understand well.

The shift to DER ensures that optimal resourcing is enabled on the grid. Load will also need to be aggregated at the edge, feeder by feeder, so that the feeder becomes a pricing node in the local DSO/DSP market. Wherever power is produced, depending on the demand and supply on the distribution grid, a dynamic rate can be established, leveraging the SGOE to ensure appropriate operational planning and pricing of the resource. The DSO/DSP, which may be the utility or a third party, in coordination with the ISO, will need to calculate dynamic pricing by district and by node, down to the smallest possible island, to optimize the production and use it most efficiently at the feeder level. The ISO will focus on large generation and coordination of multiple DSOs/DSPs for regional system planning. In other cases, the DSO/DSP may conduct its own balancing and maintain its commitment to the ISO more independently.

eMUDs.   The concept of a MUD is well-known as a means to provide water and wastewater infrastructure where city systems are not available. State legislatures empower local entities with a publicly elected board the rights to issue bonds, levy taxes, and charge utility rates in order to create infrastructure and deliver valuable services. Over time, some MUDs have expanded their purview to include other services such as garbage collection.

Given the advances of DER, it is not a far stretch to consider the use of an eMUD to provide a local community with new options for energy. As discussed in this chapter, other advances such as resource islanding may likely increase the likelihood of an eMUD emerging in the near term.

Use Case 7.3: Microgrids, Integrated ES, and Packet Power—BEMS in Action in the Future
12 a.m.–8 a.m.: Nighttime is the major off-peak cycle in any utility service territory, when electricity rates are the lowest. Small commercial businesses have a variety of strategies to take advantage of lower electricity prices, and BEMS (see Figure 7.3) enables those strategies—even automates many of them. Precooling, a tremendous energy saver, is one of the best. The air conditioner or, for larger businesses, the chiller, integrated with on-site energy production and storage, precools the office/store/warehouse/worksite according to the cooling algorithms that the BEMS learned in coordination with the business manager in the first month of operation (larger businesses will have a dedicated energy manager). Charging of forklifts and other battery-driven equipment—a category that now includes PEVs—is best accomplished overnight to avoid the high spikes in energy consumption that can prove costly under electricity rates that include a "demand" charge.

Many small commercial businesses were early adopters of the new EVs when they became available starting in 2012. Thanks to BEMS plugs and power strips, transformers can stay plugged in but no longer draw power after completing their charge (a little-known fact is that transformers used to charge devices that run on DC power continue to draw AC power, even after the devices are fully charged). Thanks to data feedback from the BEMS and its online dashboard, Ms. Small Commercial (SC) decided to have her outdoor lighting changed to LED lighting systems to lower consumption. Ms. SC is now much more aware of how much electricity each aspect of her business consumes; she even has a complete strategy for managing her businesses energy costs—something that wasn't really feasible before the advent of BEMS and its data feedback cycle [22].

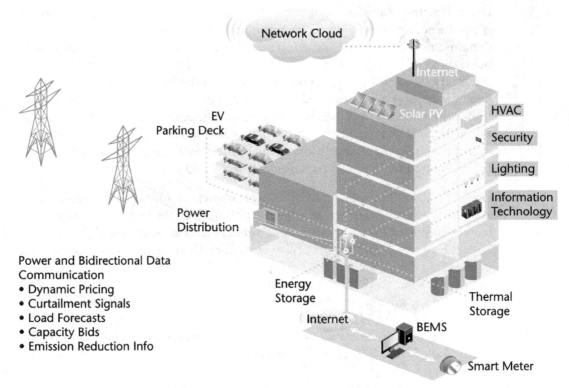

Figure 7.3 Building-to-grid (B2G) design including V2G, ES2G, DG2G, and smart devices.

8 a.m.–6 p.m.: First to arrive, Ms. SC still opens up shop and goes through her daily routine, but her routine has shifted since she became a BEMS customer and grew more aware of her consumption and, increasingly, of the potential for conservation. She no longer walks through the building, turning on all the lights and lowering the thermostat setting to get the building cooled off after it warmed up overnight. The building was already precooled according to a predetermined electricity conservation plan and lower overnight electricity rates. BEMS plugs and breakers are programmed to turn lights on and off according to a schedule based on room occupancy—"that's one less chore to start the day," Ms. SC thought, appreciating the extra time she now had.

Instead, she went straight for the PC to log in to her personalized BEMS dashboard and prepare for her weekly staff meeting at 9 a.m. As she began to scan the different screens to track the performance of her business to its preset electricity goals, she couldn't help but recall how much things had changed in the six months since she began taking control of her business electricity use. Always on the lookout to improve the bottom line, it seems that she had made a habit out of talking up electricity costs at staff meetings. She always needed to remind her staff to turn off lights and try to save electricity, but nobody ever seemed to have the same sense of urgency she did as a small business owner. Despite her best efforts, it was common in the pre-BEMS days for staff to have a running battle over the room temperature, fiddling with the thermostat despite office rules that clearly outlined office policy and required an energy-saving setting of 78 degrees. Now with BEMS automation protocols and recommended precooling strategies, the office staff has reached a general accommodation, and those arguments have become a thing of the past. Who knew saving energy could result in increased workplace harmony [23]?

Back to her work on the BEMS dashboard, Ms. SC was most concerned about office progress to their monthly consumption goals, although with each month's progress, she could see how changed behaviors were, allowing ever tighter goal setting and electricity cost reduction. After just six months, they had already dropped their electricity bill by 30%, progress that had eluded her before she had BEMS working in the background on her behalf. She printed several screenshots showing the business energy performance, including the resulting reduction in greenhouse gas emissions. She planned to share those with her staff, reflecting how much more pleasant it was to congratulate them on progress rather than harangue them on the need to save, and she knew that, like her customers, her employees were increasingly more concerned about "green" issues. She had already incorporated the good news about her new conservation-oriented workplace in the latest marketing collateral—anything to get an edge on the competition—and the new solar panels had just gone up last month, which, with their integrated ES unit, promised even more efficient operations. Already, the Sun was inching up in the sky, and the dial on the computer screen was whirling away, now and again running counterclockwise as the numbers ticked down, not up [24].

A crazy thought crossed her mind—one could even say that her business had added a new profit center, given that she was using her rooftop to generate electricity. On a whim, she jumped over to the Web site of that outfit in Dallas that was selling microturbines to fit on the eaves of buildings and capture the upflow of wind to generate electricity. Although still skeptical of how much energy she could actually generate that way, she had become more aware since getting all the data feedback on consumption that every little bit counts when it comes to electricity savings. Not only do little changes add up to big savings, but highly visible conservation—like solar panels and microturbines—also send a strong signal to her customers and prospects, even becoming another part of her marketing strategy [25]. With the advent of P2PET, a new program the utility had introduced, she realized that she could indeed start exporting her savings. She imagined a program where her employees could access free power as an employee incentive. Thanks to BEMS and other advances, she had newfound confidence to make complex electricity decisions about her business.

After the staff meeting—"that went well!" she thought—she had an appointment with a representative from the electric utility to discuss a new rate program for early adopters like herself, small commercial users now experienced with the BEMS system. Acting on her behalf, BEMS had matched her business load profile with other similar businesses and prompted the utility to make an invitation to the group of like-minded users to join a new load consolidation program. By agreeing to work in unison with the other businesses, jointly cooperating with the electric utility to lower peak demand to avoid expensive electricity production or purchase, her business and others in her group would qualify for a new rate class that was the equivalent of a large commercial ratepayer [26]. Thanks to six months working with the BEMS dashboard and the equipment in her office, Ms. SC had become well aware of her own load profile, how it compared to national averages for businesses of her type and size, and the impacts of high demand during peak times. Working independently, she had already managed to make the necessary changes in behavior to voluntarily lower her peak consumption and avoid demand charges, but now it looked like she would have a new opportunity to actually save significantly more money for doing what now came naturally to her and her staff. Next week, she would get to meet the other businesses in her new group—they planned to meet once a month to compare notes and best practices and encourage each other to save even more [27].

As she headed to lunch, she glanced at the BEMS indicator light mounted over the door, reassured that it glowed green. She knew that the adjusted goals would move it to yellow as the end of the month came near, giving her and her staff that much more incentive to be more mindful of consumption in order to meet their monthly goals. At lunch, she compared notes with a friend on the chamber board, who wanted to know more about the BEMS approach. She was surprised at the focus of the conversation, which turned more frequently to the new carbon credit market. She knew that her business had not only lowered its electricity bill but had also been responsible for eliminating tons of CO_2 in just six months of electricity conservation—that was evident from the BEMS dashboard carbon tracker—but she had overlooked the economic potential of trading in new carbon credits. Back at the office, she clicked through and found a BEMS program to manage those credits on her behalf, pooling with other BEMS users to get optimal market value [29]. "How had I managed to overlook that?" she wondered. "But so much was different in just six months, and they had come so far," she reassured herself. She made a note to investigate this new revenue opportunity.

6–10 p.m.: First to show up, last to leave, such was the life of the small business owner. Ms. SC took one last look at the BEMS dashboard—after all, she could easily check it at home if there was anything she'd forgotten—then went through her new office shutdown routine. It didn't take nearly as long to shut down the office by clicking through computer screens as it did to walk around the building and inspect light switches, systems, and so forth. No longer strictly reliant on a mental checklist, now she merely had to quickly review dashboard screens that monitored the BEMS, thermostat, breakers, and plugs, detailing current operational status as well as up-to-the-minute energy consumption levels, comparing them to preset goals based on best practices and industry norms. Not only did it take less time to manage her electricity consumption, it was far more effective than the old system of manual checks and balances, individual smart thermostats, and other utility efficiency programs. Thanks to BEMS, electric expenses had been transformed from one of the hardest-to-manage line items on her income statement to one of the easiest. In fact, she pondered what life would be like if there were a system like BEMS to help her manage all the other items on her list.

Integrated ES, VPPs and Microgrids

Electrical ES at utility scale remains both the Holy Grail for electric industry redesign and one of the most vexing energy technology challenges. Storage applications may include: peak shaving (load leveling) systems to help commercial and industrial users manage electricity costs under variable utility tariffs and to help utilities manage generating assets to minimize waste; renewable integration systems to help power producers, utilities, and end users cope with the inherent variability of wind and solar power, transforming it into firm, dispatchable power, and to better match peak wind and solar output with peak demand; power quality systems to protect commercial and industrial users from interruptions that cost an estimated $75–200 billion per year in lost time, lost commerce, and damage to equipment; and transmission and distribution support systems to help utilities reduce grid congestion, defer upgrades, and minimize waste.

Because utilities still depend on spinning reserve and other supply-side strategies to ensure reliability—and even do long-term planning on the assumption that ES will not be viable in the foreseeable future—adding ES into the system will significantly disrupt the current energy ecosystem. Implementing an advanced smart grid will help utilities to develop a vision based on the potential of affordable ES, where they will move from R&D to active trials of various technologies in different parts of their service territory.

The use of ES does not need to concern any loss of comfort or convenience, but rather to accept some minor sharing of the resource in a collective strategy to pull in a new resource that has never existed before. An advanced smart grid will enable solutions that leverage DER, tapping into some that haven't yet been conceived. Integration of ES with other assets, including DG, DR, EVs, VPPs, microgrids, and smart grid technologies will be vital to maximizing the value of each individual new resource, but also of the collective advanced smart grid. With a network and SGOE, all these distributed elements can be optimized and their potential realized.

Challenges in creating a utility ES program from a strategic perspective will include not only ES integration, but also designing the system and prioritizing ES locations. ES is likely to be located where there is congestion on the grid.

The role of ES in a disaster recovery situation will need to be considered. ES will be valuable in restoration of electricity service after a massive outage. Prioritizing DG and ES for disaster shelters will buy utilities time, since there will be a minimal amount of power assured in those spots in the case of a prolonged outage. Colocation of DER, specifically DG and ES, with disaster shelters will support utility goals in a disaster recovery plan. VPPs describe a demand-side alternative to accommodate growth in peak demand to the traditional supply-side alternative of adding a natural gas power plant, commonly referred to as a *peaking unit* or a *peaker*.

At the microlevel, VPPs will require technology to be refined at the scale of a single distribution feeder or neighborhood.

At the macrolevel, integrating a VPP system to the grid will require the ability to manage new levels of complexity in remote sensing, control, and dispatch. Integration of the complex VPP systems will require the utility control center to be able to "see" the status and availability of such distributed capacity. The complexity of system dispatch with these types of resources will require automated decision-making to signal a direct load curtailment condition to DR resources to make optimal use of these new resources and integrate into the larger system portfolio.

New control software now available in the market is designed to enable utilities to manage the numerous and complex dispatch requirements of VPPs. Intelligent edge devices such as smart meters, smart routers, and smart inverters are now capable of communicating their operational status, calculating the ramifications of their actions on their surrounding environment and making decisions to change their state in real time so that the network becomes self-healing and self-adapting. Autonomous, edge-based decision-making maintains safe energy flows, minimizes service disruptions, and, perhaps most importantly, helps to avoid catastrophic damage. Such a distributed network of smart devices connected to VPPs will provide them the intelligence they need to create new demand action capabilities that integrate new edge-based resources seamlessly with the advanced smart grid.

VPPs and microgrids represent an emerging resource and application of creative bundling of multiple technologies, rapidly becoming a key aspect of energy policy going forward. Connecticut, for instance, has written microgrids into law and is actively promoting their creation. The DOD is active with its Smart Power Infrastructure Demonstration for Energy Reliability and Security (SPIDERS) program to study microgrids for military facilities. Utilities may own microgrids themselves—some are considering doing so for purposes of grid reliability; third parties may own and operate microgrids on behalf of property owners, or property owners may approach microgrids themselves without outside support.

The concept of a microgrid—a self-contained, multi-resource, multi-load, and islandable energy entity—may be applied to other areas besides a business or a campus, as used in our example. A microgrid in a single building might be referred to as a nanogrid, enabling a building to connect and disconnect from the grid at will. Imagine the DR capabilities when buildings are so enabled. Similarly, a microgrid may be used as a new way of organizing a single substation distribution utility as is found in many small MOU systems.

New concepts and thinking about how a grid works will be needed for these approaches to take hold. For instance, while initial adopters of these solutions may have off-grid operations in mind (step 1 could be called "disconnect"), the actual realization of these concepts will be difficult. Step 2 could be called "independent operation," and it will borrow from our knowledge of how we operate the grid today and concepts in this book about future operational protocols. Step 3 is likely to concern the previous discussion on resource islanding, with a focus on "reconnection for reliability." Step 4, which we could call "replacement," will finally be realized as renewable energy and DER are widely adopted as policy. As VPPs and microgrids mature, they will become the LANs and WANs in the SG3 world, connected by an energy Internet and offering new levels of independence and interdependence that we have barely contemplated today.

The twenty-first century demands a new set of organizing principles when it comes to electricity, suggesting a complete business transformation as outlined in Chapter 6. We have experienced these changes on multiple dimensions, as shown in Figure 7.4, starting at noon (using a clock metaphor) and moving clockwise around this circle of progress of the twentieth-century grid fundamentals and the sea change to a new set of twenty-first-century grid-guiding principles:

- Where the early grid developers embraced access to cheap, plentiful fossil fuels like coal, petroleum, and natural gas, building ever larger, more efficient coal-fired power plants, today we're challenged to avoid the carbon that results from burning fossil fuels.
- Where grid resources on the supply side expanded to meet a growing population and increasing use of electricity, today we're challenged to involve consumers to avoid peak load by making better use of existing and new types of DER.

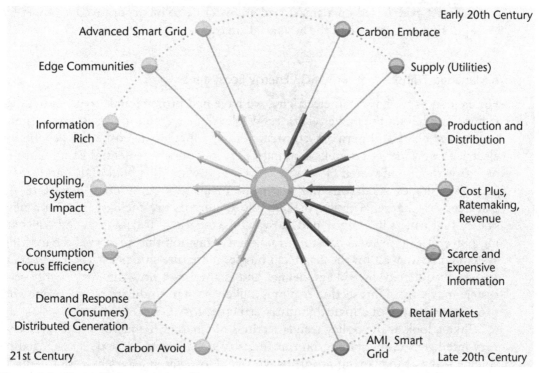

Figure 7.4 Electricity economy transition.

- Where production and distribution of electricity were central to planning and building the grid, today we also focus on our built infrastructure to use energy more efficiently and to better understand energy consumption behavior patterns.

- Using cost plus ratemaking to establish revenue targets for regulated utilities, the keystone of the twentieth-century "regulatory compact" that allowed monopoly franchises, is giving way in today's more dynamic environment to new pricing calculation considerations that include decoupling and more frequent true-ups to continue to equitably balance the need to meet customers' needs with those of the stakeholders.

- When information used to be scarce and expensive, utilities devised ingenious, artful methods to plan and operate the complex grid, but in today's environment of abundant, low-cost information, we are challenged to add sensors to gather ever more data and then use data analytics to plan, operate, and manage complexity.

- Late in the twentieth century, retail competition showed faltering progress and potential, and while markets still have a prominent role to play, in the twenty-first century communities out on the edge will leverage social networking, smart mobile devices, and other lessons learned from the Internet to identify and deliver greater value.

- Finally, where AMR evolved into AMI and led the charge to deliver a smart grid, in the twenty-first century we will need a more expansive definition of smart grid, based on a smart grid architecture to incorporate edge power by design, which we call the advanced smart grid.

A Managed Transition and the SG3 Energy Ecosystem

For as long as we have had electricity, we have had professionals to generate, distribute, and bill for the power via rates. We developed a monopoly system to stimulate power plant and network construction and spread the costs and benefits of electricity as wide as possible, making kilowatt-hours an essential of human life on par with air and water. However, as we've shown in this book, DER and smart grid technologies are driving our system to a transformation into a new state. As discussed in Chapter 6, policymakers and regulators are racing to design a new system that ushers in a more modern way of generating, distributing, and consuming power. Utility leaders call for a managed transition that preserves the existing system while we wait for the new technologies to mature sufficiently. SG2 and SG3, as described herein, provide guidelines and examples of how this new energy ecosystem may work. This section explores a likely path for industry restructuring and provides a glimpse of a logical industry structure for SG3.

Take a look at the policy trends sections of Chapter 6 and you see a variety of very intelligent folks opining on our industry's challenges, or actively engaged in solving some of the vexing problems we face. However, it leaves us with the question: Where does this industry go from here? As we look at the emergence of SG3, which will arise with the implementation of the ideas in this book and maturing trends, it is worthwhile here in Chapter 7 to speculate for a moment on a likely outcome.

One hundred years ago, in the early part of the twentieth century, Samuel Insull and others developed and launched the business model that became our modern electric utility industry. At the time, there were regions in the country where a utility pole might carry multiple distribution lines that overlaid each other. The argument for a monopoly on the wires side made a lot of sense. Soon, electric utilities were trading away private sector rights like pricing and market independence in exchange for rights to service all the customers in a prescribed service territory. Samuel Insull reasoned that a monopoly would accelerate improvements in generation technologies, and indeed, George Westinghouse famously worked with Insull to meet his needs at Commonwealth Edison in Chicago as more and more customers bought into electrification and bought new appliances.

In addition to large cities like Chicago, the business model basics of monopoly grant, regulatory compact, and affordable kilowatt-hours for everyone took shape in the Midwest under Insull's Midwest Utility Holding Company. Concentrated ownership by a holding company of multiple geographic service companies was another feature in the early years. FDR's attorney general addressed this challenge by making Samuel Insull public enemy number one and the personification of excess that electric utility industry had become. PUHCA of 1935 brought the reforms needed, and the industry had a good long run under the watchful eyes of regulators. What now?

Almost 100 years later, as we contemplate the future, we can speculate that electric utilities will no longer be alone in providing electricity. DER represents an emerging alternative for grid-delivered electricity. We already see competition emerging from companies like Solar City, Vivint, and others, which leverage PPAs to finance long-term contracts for electricity service from on-site solar PV systems, and soon, paired with ES. As we can see, utilities will remain critical elements of our modern economies well into the future. However, new serious competition for edge power infrastructure and services is growing at a rapid pace globally, making utilities no longer the only game in town. The nodes at the ends of the links—the buildings where consumers are—have joined the system as more active participants in the energy ecosystem. Transformed infrastructure companies will need to make room for new energy players that offer value-added services associated with their competitive advantage and core competencies.

Let's put a stake in the sand then and make some predictions, in two general categories: infrastructure and value-added services.

Infrastructure

First, we can expect to see infrastructure converge on three basic models globally. Alternatives 1–3 in the following focus on the transformation of the electric utility power companies—the grid operators as we know them today—and the resultant impact on infrastructure operations.

1. *Investor-owned utility consolidates:* We should expect the grid to continue to operate and enable valuable services, but we should not expect the historic diversity of utilities to continue. Consolidation of IOUs will see the strong shareholder-owned utilities buying up the weak, as disruption grows and those who fail to craft and implement successful sustainability plans become acquisition targets of those who do. We see hints of this over the past several years with the emergence of even larger electric utility companies like Duke Energy, Exelon, NRG, Nextera Energy, and even Berkshire Hathaway Energy. Operational efficiencies in the large merged entities will help the larger utilities to buy time and figure out successful new business models. There will be fewer IOUs, but they will be larger, healthier, and more focused on transformation, competition, and long-term sustainability.

2. *Community energy emerges:* Likewise, the raft of smaller member owned electric cooperatives (coops) are likely to find efficiencies by consolidating their operations and eliminating redundant positions—at a minimum, we may see them unite in loose federations to share costs. As this consolidation progresses, a new model of utility operations will emerge that provides local communities more say in the grid operations and decision-making. With a maturing consumer base, the concept of *community energy* will emerge as an alternative to shareholder-owned electric utilities (IOUs). Keeping revenues inside a local community and reinvesting profits will lead to greater proliferation of clean energy alternatives, more efficient operations, and energy independence that becomes a differentiator to drive regional economic development.

3. *Public power gets lean:* Public power will follow a similar pattern, as government-owned electric utilities reflect the will of voters, who gain a voice

as consumer maturity progresses. Around the world, where federal, regional or local governments (MOUs) own and operate utilities, energy will become increasingly political. Bureaucracies accustomed to slow change will be challenged by the imperative of transformation, whether from pressures to adopt new energy technologies, or from increasing rates, or from increasing subsidies from governments to keep insolvent utilities afloat. Public power is a strong concept, powerful in cultures where it has long been the norm. We can expect *smart cities* to converge with *smart grid* in countries where public power is dominant. Here, converged infrastructure offers keen advantages enabled by new technologies.

Value-Added Services

The twenty-first-century economy depends on value-added services that combine innovative technologies and business models. New DER technologies in particular offer a variety of new tools to personalize energy and change the fundamental value proposition of electricity. We're already seeing some early innovators in areas where electricity sales have been deregulated, among REPs and IPPs who are stepping out from the traditional fold to add services to their power sales. Likewise, we can expect to see other industries adopt new DER capabilities to enhance their own core competencies and core business strategies, leveraging their strong brands with consumers by offering appealing, innovative approaches to electricity shaped by their distinctive perspectives.

4. *REPs:* This is an area ripe for the addition of DER and new services to complement a retail energy contract and keep energy customers home on the range. When all you sell is a kilowatt-hour, there is always a new, lower price beckoning your customers to switch to a different provider. *Customer churn* is a particular challenge in the deregulated retail electricity industry. Watch for special pricing to create new value-added services, including flat rates and bundled solar PV. These new services will be used to extend contract lives and lock in customers for longer periods, not unlike the way that smart phone discounts are used in the telecom space to extend the network service contract an additional two years.

5. *IPPs: Whither central generation? What is to become of central power plants as more and more DER comes onto the market?* No doubt, older, dirtier units will be retired and decommissioned. However, other plants with cleaner operations and long lives ahead of them will retain positive value, if not their book value. Large or specialized IPPs are likely to acquire distressed assets to add to generation fleets that also include renewable energy farms. These assets will serve power-hungry industries that require more intensive energy or special approaches to grid energy not offered by utilities. Expect to see specialist IPPs that sell directly to large customers with large or unusual power needs in bilateral contracts, as new regulations are put in place, and monopolies begin to lose their exclusive grip on power sales.

6. *Large industrial and commercial customers:* Now we shift to energy consumers who will find new ways to become prosumers, as technology matures and as their outlook shifts with new perspectives. Large industrials and commercial customers have a need for large amounts of power, highly reliable power, and/or high-quality power—or some combination thereof. They view electricity as an input to their business model, and so, like IPPs, large factories, foundries, and server farms, may choose to purchase devalued generation assets to secure their own power needs, selling excess production into wholesale markets, or even building new energy businesses marketing long-term PPAs. Chapter 6 details some server farms that are already innovating with renewable energy and DER. Large property developers are already quite active in exploring the potential of microgrids, and this can only accelerate in the future.

7. *Small commercial customers (retail companies):* Small commercials and, specifically, retail companies have at least two angles on DER to complement their core competency, selling retail products and services. Big-box retailers like Home Depot, Lowe's, IKEA, Costco, and Walmart have already entered the energy space in a natural way, by selling retail DER products, from LEDs to solar PV cells, and by bundling services with local companies. Walmart, in particular, has long been a vocal supporter of the emerging DER industry, and we can expect to see retailers like Walmart and others using DER more and more to green up their image, lower their energy costs (especially demand charges), and increase service reliability. Small commercial companies have long had to pay high demand charges based on their load shapes; with DER, they will have new options to lower their bills, and as they grow more familiar with DER, its easy to imagine some adding innovative DER services to their value portfolios.

8. *Small commercial customers (commercial real estate owners and property managers):* The core competency in this industry is based on managing, owning, or buying and selling real estate. Traditionally, these businesses have seen energy as a cost to manage with BEMS and BASs. Now, on-site energy represents a huge opportunity to transform commodity real estate into PEBs to launch a new class of real estate, demanding adjustment of cap rates and comparable values. In this space, expect to see tremendous innovation leveraging such technologies as VPPs (DG, BEMS, ES, DR), PEBs, microgrids, and nanogrids. Landlords selling energy directly to tenants, currently a REP function, represents another innovation that will leverage DER. It is worth noting that the State of Texas passed House Bill #2049 in September of 2013 as the nation's first law allowing building owners in the competitive electric territories to become utility providers for their tenants after installing DER capabilities. This is clearly a game changer that will be replicated by the other 49 states soon enough.

9. *Transportation companies:* The core competency of transportation companies is to move people and goods from one place to another. Those in the transportation business are most likely to look at stored energy as the key innovation. Electrification of the transportation sector may hold the most potential of all, as climate change and advancing DER technologies lead

more and more to consider an EV for personal transportation, as fleets migrate to an EV platform, and as new service programs like Uber and Lyft emerge. Further innovations are highlighted in this chapter, including V2G and EV cooperative programs, which will offer collaboration opportunities with infrastructure companies. In the meantime, Tesla has become the model of innovation in the transportation sector, with charging networks that offer free charging for the lifetime of a Tesla owner, with EV batteries integrated with PV systems at Solar City, and with batteries set to realize their tremendous potential as they come down in price, spurred on by technology advances and economies of scale from increased production levels (i.e., Tesla's Giga Factory). Because EVs subsidize ES, they are a natural for disruptive innovation in the new value-added services sector. EVs hold the most potential to introduce ES into SG3.

10. *Entertainment/telecom/Internet companies:* In fits and starts, the HEMS and HAN space has grown with better and better technology. With a core competency of offering entertainment, communication, and security services to residential and business customers, companies like ATT, Verizon, Time Warner, Google, and Apple have long eyed the electric industry with a glint in their collective eye, but they have had limited success in penetrating this insular industry. This sector is attractive to communication and technology infrastructure companies not only because it offers a natural extension of their *triple play,* but also because these brands are so strong with consumers, they are well positioned inside the home and have a leg up on developing and offering value-added services when compared to their electric network services cousins, the electric utilities. The direction of innovation in this sector is likely to include energy management, but really all flavors of DER are in play, leading to the proliferation of PEBs and the realization of the vision of transactive energy. A variety of platforms have been established or are being planned by these companies, and collaboration with utilities is highly likely.

11. *Energy entrepreneurs:* The field of entrepreneurial start-ups is positioned to leverage new technological capabilities as they emerge and carve out value-added niches solving problems that were heretofore unsolvable. Look for these small companies to grow rapidly and to be snapped up by the established players listed above. With a premium on innovation and a rapidly changing technology landscape, energy entrepreneurialism should dramatically expand in the years ahead.

12. *Residential consumers and prosumers:* Residential consumers will increasingly participate in these diverse new retail service models now under way as a key market. They will buy the emerging new value-added services, and they will invest in DER equipment and services to begin to produce electricity on their own. Moreover, as consumers become more mature, they put themselves on the path to becoming prosumers, and then to a more innovative and independent relationship with the utilities that serve them. Innovations may center upon participation in new community energy business models discussed at the start of this section.

13. *Electric distribution utilities:* Finally, we arrive back where we started, with the infrastructure companies. As the operators of distribution grids that extend out to the edge of the networks, where they integrate with DER, electric utilities are in fact well positioned to innovate by developing new value-added DER products and services. However, regulatory rules too often inhibit utilities by discouraging risk taking. Further, utility cultures generally still value consistency over innovation. Still, we see many utilities already focused on developing products and services using DER; some of these will succeed and develop this new core competency, while many more will find such changes tough sledding, ultimately opting to partner with the companies in this list to take joint programs to market.

The Advanced Smart Grid: Edge Power Driving Sustainability

The advanced smart grid offers the utility industry and related stakeholders a new framework and operational paradigm for energy creation, distribution, and consumption, but also a transformational social and transactional platform.

Our work and play—our very lives—will be transformed as the advanced smart grid becomes pervasive. It has been said that the smart grid will exceed the impact of the Internet, and we would not dispute that assertion. The advanced smart grid will accelerate job creation, and, more importantly, it will stimulate the emergence of edge power and foster a new age of renewable energy and DER. The advanced smart grid will enable end-to-end cybersecurity from the device through the network to the core of the utility and back.

The advanced smart grid will enable a transition from the incredible complexity of today's grids to an enhanced simplicity of use by the utility. The advanced smart grid will be built with an open standards–based SGAF to support data dissemination on TVs, smart phones, tablets, and computers. The advanced smart grid will not be about any single network technology but rather the integration of multiple IP networking technologies using a single SGOE. The advanced smart grid will stimulate data analytics and new transactions to engage utilities, customers, and independent energy producers in a new, self-healing, interactive energy ecosystem of energy creation, distribution, and consumption reaching out to millions of smart edge devices in the emerging IoT.

The advanced smart grid design, as described in Figure 7.5, will become the manifestation of a new horizontal energy ecosystem that replaces utility silos across the four existing domains: generation, transmission, distribution, and metering services, as well as the four emerging domains of DER: DR, DG, ES, and EVs. The advanced smart grid design will integrate the utility with end consumers in a variety of new interfaces, including B2G, H2G, V2G, ES2G, and DG2G.

The advanced smart grid will introduce us to new concepts and capabilities and new ways of thinking about energy and how it impacts our lives. The advanced smart grid will enable a new social and transactional platform for clean and abundant power.

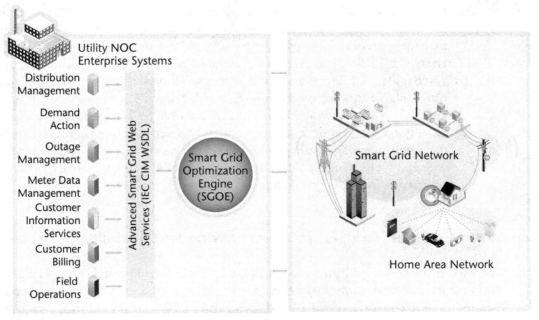

Figure 7.5 Advanced smartgrid design.

Endnotes

[1]   The vaudeville act on The Ed Sullivan Show where a professional plate spinner lined up a series of pool cues and proceeds to keep increasingly more plates spinning on them is an excellent example of what grid operators will have to contend with when DER blossoms; see it on YouTube at http://www.youtube.com/watch?v=ZhoosIoY404.

[2]   Special charging rates discourage rapid charging that stresses the utility system.

[3]   Special charging rates encourage usage during nonpeak hours, when renewable wind energy is abundant.

[4]   Smart EV charging system is a new smart end device equipped with intelligence.

[5]   EV charging becomes a new revenue source for utilities, which control location to accommodate grid acceptance of EV charging.

[6]   Incentives guide drivers to charging location sections that are conducive to a healthy grid.

[7]   Account correlation with mobile charging enabled by advanced smart grid.

[8]   Combination of solar PV and EV enabled by smart inverters.

[9]   Predictive volt VAR program and two-way power flow use EVs to help with balancing.

[10] Digital billing draws usage information from a common database.

[11] Two-way information flow enables "mobile" rates that are driven by usage and account, irrespective of location.

[12] Prechilling the home is an example of thermal energy storage as a DR strategy.

[13] A good example of a strategy to address the challenge of "vampire power."

[14] Multiple screens will provide access to HEMS information and remote control for ultimate ease of use and optimal fine tuning and control.

[15] A new, more expansive definition of a "smart" thermostat.

[16] Pool pumps consume more energy than almost any other appliance in the home.

[17] Tying conservation goals and motivations such as climate change to data feedback is a powerful means of making changes in energy consumption behaviors to be permanent adjustments and new habits.

[18]   Information on distributed generation is integrated with information on demand response and EVs.

[19]   Providing a check on utility bills is a benefit that doesn't get much play when HEMS is discussed.

[20]   HEMS tools and data feedback has brought the family through the cycle to prosumers.

[21]   This is an example of aggregated thermal storage to provide DR during an extreme peaking event.

[22]   Small commercial customers are likely to be at the front of the line as energy consumers become empowered.

[23]   Information and control provide significant empowerment and reduce distractions.

[24]   Integrated ES, DR, and DG.

[25]   Clean energy behaviors will become a more powerful marketing tool to differentiate businesses as consumer empowerment grows and options for distributed energy proliferate.

[26]   A likely new service for utilities will be to provide aggregation services to consolidate load and encourage more collective behavior.

[27]   This is a good example of a virtuous BPI cycle described in Chapter 2.

[28]   Green tag and white tag markets will develop to pool renewable energy and energy efficiency credits.

Acronyms and Abbreviations

3G	Third-generation wireless telecommunication network
4C	Comfort, convenience, cost, and carbon
4G	Fourth-generation wireless telecommunication network
ADR	Automated demand response
ADS	Association for Demand Response and Smart Grid
AGC	Automatic generation controllers
AGI	Advanced grid infrastructure
AMI	Advanced meter infrastructure
AMR	Automatic meter reading
AMS	Asset management system
APPA	American Public Power Association
ARRA	American Recovery and Reinvestment Act of 2009
AWEA	American Wind Energy Association
BAS	Building automation system
BEMS	Building energy management systems
BOS	Balance of system
BPI	Business process improvement
BPL	Broadband over power line
CAIDI	Customer average interruption duration index
CCET	Center for the Commercialization of Electric Technologies
CEIDS	Consortium for Electricity Infrastructure to Support a Digital Society
CES	Community energy storage
CFL	Compact fluorescent lighting
CHP	Combined heat and power
CIP	Common Internet protocol
CLEC	Competitive local exchange provider
CO_2	Carbon Dioxide
CPP	Critical peak pricing

CSAF	Current state architectural framework
DA	Distribution automation
DCS	Distributed control system
DEMS	Distributed energy management system
DER	Distributed energy resources
DEROE	DER optimization engine
DERMS	Distributed energy resource management system
DEWG	Domain expert working group
DFA	Distribution feeder automation
DG	Distributed generation
DG2G	Distribution grid to grid
DIPP	Distributed independent power producer
DITL	Day-in-the-life (scenario)
DMS	Distribution management system
DNP3	Distributed network protocol, version 3
DOE	Department of Energy
DPS	Digital premise server
DR	Demand response
DRCC	Demand Response Coordinating Committee
DRMS	Demand response management system
DRSGC	Demand Response Smart Grid Coalition
DSO	Distribution system operator
DSP	Distribution service platform provider
DSPV	Distributed solar photovoltaic
DVD	Digital video disk
EDF	Environmental Defense Fund
EE	Energy efficiency
EEI	Edison Electric Institute
EISA	Energy Independence and Security Act of 2009
EMS/SCADA	Energy management systems and supervisory control and data acquisition
eMUD	Energy municipal utility district
EPRI	Electric Power Research Institute
ERCOT	Electric Reliability Council of Texas
ES	Energy storage
ESA	Energy Storage Association
ETS	Energy thermal system
EV	Electric vehicle
EVSE	Electric vehicle support equipment

FCC	Federal Communications Commission
FDIR	Fault detection, isolation, and restoration
FERC	Federal Energy Regulatory Commission
FIT	Feed-in tariff
FOA	Funding opportunity announcement
GAO	Government Accounting Office
GIS	Geospatial information system
GMS	Generation management system
GWA	GridWise Alliance
GWAC	GridWise Architecture Council
H2G	Home-to-grid
HECO	Hawaii Electric Company
HEMS	Home energy management system
HERS	Home energy rating system
HPPV	High-penetration PV
HVAC systems	Heating, ventilation, air conditioning systems
ICE	Internal combustion engine
ICT	Information and communication technology
IEC	International Electrotechnical Commission
IEEE	Institute of Electrical and Electronics Engineers
IHD	In-home displays
ILM	Integrated load management
ILSR	Institute for Local Self-Reliance
IoT	Internet of things
IOU	Investor-owned utility
IP	Internet protocol
IPP	Independent power producer
IRP	Integrated resource planning
ISO	International Standards Organization
ISSGC	Illinois Statewide Smart Grid Collaborative
IT	Information technology
ITIL	Information Technology Infrastructure Library
ITT	Information technology and telecommunications
KPI	Key performance indicator
kW	Kilowatt
kWh	Kilowatt-hour
LACE	Levelized avoided cost of energy
LANL	Los Alamos National Laboratory
LBNL	Lawrence Berkley National Laboratory

LCOE	Levelized cost of energy
LED	Light Emitting Diode
LEED	Leadership in Energy and Environmental Design
LFG	Landfill gas
LP	Long-playing vinyl record
LPPC	Large Public Power Council
LTC	Load tap changer
LTE	Long-term evolution
M2M	Machine-to-machine
MBO	Management by objective
MCC	Microelectronics & Computer Technology Corporation
MDM	Meter data management
MDMS	Meter data management system
MDU	Multidwelling unit
MOU	Municipally owned utility
MUD	Municipal utility district
MW	Megawatt
MWh	Megawatt-hour
NARUC	National Association of Regulatory Utility Commissioners
NASUCA	National Association of State Utility Consumer Advocates
NETL	National Energy Technology Laboratory
NERC	North American Reliability Council
NIST	National Institute of Standards and Technology
NOC	Network operations center
NRECA	National Rural Electric Cooperative Association
NREL	National Renewable Energy Labs
NRTC	National Rural Telecommunications Council
OCPP	Open charge point protocol
ODM	Original device manufacturer
OEM	Original equipment manufacturer
OLA	Operational level agreements
OMS	Outage management system
OSI	Open systems interconnection
OSTP	Office of Science and Technology
OT	Operations technology
P2P	Peer-to-peer
PAP	Priority action plan
P&L	Profit and loss statement
PEB	Positive energy building

PEM	Positive energy mortgage
PHEV	Plug-in hybrid electric vehicle
PLC	Programmable logic controller; also Power line carrier
PMI	Project Management Institute
PMO	Project management office
PMU	Phasor management units
PPA	Power purchase agreement
PSAF	Pecan Street Architecture Framework
PSC	Public Service Commission
PUC	Public Utility Commission
PUHCA	Public Utility Holding Company Act
PV	Photovoltaic
PVCC	Predictive volt/VAR control
QoS	Quality of service
REP	Retail energy provider
RES	Renewable energy standard
RFI	Request for information
ROI	Return on investment
RPS	Renewable portfolio standard
RTP	Real-time pricing
RTU	Remote terminal unit
RUP	Rational unified process
SA	Substation automation
SAIDI	System average interruption duration index
SAIFI	System average interruption frequency index
SCADA	Supervisory control and data acquisition
SEER	Standard energy efficiency rate
SEGIS	Solar Energy Grid Integration System
SEIA	Solar Energy Industry Association
SEPA	Solar Electric Power Association
SGAC	Smart Grid Architecture Committee
SGAF	Smart grid architecture framework
SGCC	Smart Grid Consumer Coalition
SGIP	Smart Grid Interoperability Panel
SGOE	Smart grid optimization engine
SLA	Service-level agreement
SOA	Service-oriented architecture
Solar PV	Solar Photovoltaic
TCO	Total cost of ownership

TOU	Time of use
TSW	True sine wave
USGBC	U.S. Green Building Council
UTC	Utilities Telecommunications Council
V2G	Vehicle-to-grid
V2H	Vehicle-to-home
VOIP	Voice over IP
VOST	Value of solar tariff
VPN	Virtual private network
VPP	Virtual power plant
WiMAX	Worldwide Interoperability for Microwave Access
WMS	Workflow management system

Glossary

The smart grid is a challenging field for many reasons, not the least of which is that it draws on multiple disciplines. To understand smart grid requires a basic understanding of terms from power engineering, telecommunications, IT, business and economics, and public policy, including legislative, regulatory, and stakeholder terms and groups. For the reader's convenience, we provide this glossary and list the chapter where the term is first mentioned. Terms created to capture new concepts and approaches related to the advanced smart grid are italicized and denoted with an asterisk (*).

3G (third-generation) (Chapter 2) 3G is a generation of wireless telecommunication network technology standards developed according to specifications from the International Telecommunication Union for mobile telephony. 3G must provide at least 200 Kbps, and the 3G family includes UMTS, used primarily in Europe; CDMA; and EVDO.

4C* A term we coined to reference the four parameters by which energy consumers will make energy decisions in the future: comfort, convenience, cost, and carbon. Using an HEMS device, the energy consumer will set 4C parameters, and the system will interact with home appliances and the utility to optimize personal comfort and convenience, energy cost, and carbon footprint according to the unique preferences of the energy user.

4G (fourth-generation) (Chapter 2) A successor generation to 2G and 3G, 4G wireless telecommunication network technology is the fourth generation of cellular wireless standards with speed requirements of 100 Mbps for high-mobility communication from moving vehicles, and 1 Gbps for pedestrians and stationary users. 4G will also include comprehensive, secure IP-based functionality. LTE is a synonym for 4G.

Advanced meter infrastructure (AMI) (Chapter 2) AMI is comprised of smart meter end devices, a wireless communication network, and data backhaul network, integrated to provide interval consumption data collection and processing for use in revenue metering and bill production. AMI systems also provide ancillary functionality, including outage management information and remote turn on/turn off. This component of a smart grid project involves the deployed smart meters, network connectivity (generally wireless communications, PLC or BPL between the endpoints on the network and the energy control center, and any hardware and applications that enable interoperability and automated response using dedicated or shared databases.

Advanced grid infrastructure (AGI)* AGI is a term we coined for what may succeed AMI, using real-time system information and control to integrate the functionality of advanced metering not only with demand response and outage management and restoration, but also with such DA functionality as capacitor bank control, switch control, volt/VAR control, FDIR, fault detection and isolation and restoration, distribution management, and substation management. AGI would also include integration at the DER level through inverter management, providing management of solar PV systems, EV charging systems management, and remote energy storage devices. AGI functionality will ensure that the utility remains in control of all devices connected to the advanced smart grid, addressing an emerging business risk for utilities when consumers gain increasing amounts of responsibility for edge device management.

Advanced metering infrastructure (AMI) (Chapter 3) AMI is a two-way WAN solution, unlike AMR (listed below) which is only a drive-by LAN or a one-way WAN solution. It also automatically collects the consumption, diagnostic, and status data from water meters, gas meters, and electric meters, and transfers the data to a chosen central database for billing, troubleshooting, and analyzing. This technology allows for remote connect and disconnect of the meter saving for each truck roll; it also saves the expense of periodic trips to each physical location to read a meter; it allows for billing to be based on near real-time consumption rather than on estimates based on past or predicted consumption to create new services (e.g., TOU rates), and it allows for proactive outage detection at each meter for each premise that they serve.

Advanced smart grid* (Chapter 1) The architecture and network that enables the seamless integration of central generation, transmission infrastructure, substation infrastructure, distribution infrastructure, and meters, with buildings, electric transportation, DG, ES, and smart devices to increase grid reliability, EE, renewable energy use, and customer satisfaction, while reducing capital and operating costs..

American National Standards Institute (ANSI) (Chapter 2) ANSI is a 501(c)3 organization that administers and oversees the U.S. standards and conformity assessment system. ANSI oversees the creation, promulgation, and use of norms and guidelines that directly impact businesses in nearly every sector, including equipment used in the U.S. electric utility industry. ANSI accredits the programs that assess conformance to standards—including globally recognized cross-sector programs such as the ISO 9000 (quality) and ISO 14000 (environmental) management systems.

American Public Power Association (APPA) (Chapter 6) The APPA, created in 1940, is the nonprofit association that represents the nation's public power utilities, whose membership includes over 2,000 community-owned electric utilities, which collectively provide service to over 46 million Americans. The mission of APPA is to advance the public policy interests of its members and their consumers, as well as to provide member services to support adequate, reliable electricity at a reasonable price while protecting the environment.

While some members are large utilities (e.g., LIPA, LADWP, and the Salt River Project) and the largest have their own association as well (i.e., LPPC), most members are distribution-only utilities with less than 10,000 meters.

American Recovery and Reinvestment Act of 2009 (ARRA) (Chapter 5) Better known as the stimulus bill, this legislation directed nearly $4B in matching funds to support smart grid investment grants (DOE FOA 58) and smart grid demonstration grants (DOE FOA 36).

American Wind Energy Association (AWEA) (Chapter 6) AWEA, the industry trade association representing the wind energy industry, focuses primarily on large remote wind systems. Given that this book has discussed the smart grid from the distribution grid perspective, wind energy has limited impact on the smart grid, since most wind energy flows onto the transmission grid. A BCG report found on-shore wind energy already at rough grid parity, but that integration with the grid will become an ever growing challenge, given the remote location of wind energy sites and limited transmission capacity. Still, in 2010, a demonstration project for CCET, which received $13.5 million DOE ARRA FOA 36 funding, made significant progress in its plan to tie together wind energy with offsetting DER using smart grid technology. In a unique combination, the project integrates intermittent wind energy resources on one end of the grid using synchrophasors with offsetting balancing resources at the other end of the grid, in a Houston suburb, where a VPP—a combination of distributed solar PV, energy storage and HEMS—will mitigate the otherwise disruptive impacts of intermittency from wind turbines.

Automatic generation controller (AGC) (Chapter 2) AGC is a mechanism for automated power system control and operation, with a specific purpose to balance power in the system and maintain constant system frequency. AGC acts to provide primary automatic control of generators on the grid, reacting to frequency changes, and operators can exert secondary control with corrective actions. Finally, AGC is used to execute actions to ensure economic dispatch of generation units.

Automatic meter reading (AMR) (Chapter 1) AMR is a drive-by local area network (LAN) or one-way wide area network (WAN) technology for automatically collecting consumption, diagnostic, and status data from water meters, gas meters, and electric meters, and transferring the data to a chosen central database for billing, troubleshooting, and analyzing. One-way WAN AMR saves utility providers the expense of periodic trips to each physical location to read a meter. Both drive-by LAN AMR and one-way WAN AMR enable the billing to be based on near real-time consumption rather than on estimates based on past or predicted consumption. This timely information coupled with analysis can help both utility providers and customers better control the use and production of electric energy, gas usage, or water consumption.

Backhaul (Chapter 2) Short for Internet backhaul, this term refers to the transportation of data from the wireless local area network (LAN) back to the Internet. A LAN provides local area bandwidth but also requires connection to the internet. Backhaul capacity and performance will impact LAN performance, the number of connections the LAN will support, and the data carrying capacity of the LAN.

Broadband over power line (BPL) (Chapter 2) BPL is an upgrade to power line carrier, a technology that uses existing electric utility lines to bring a wired information signal to the end user. In the case of BPL, the wired signal is broadband, running at speeds that typically range between 1.5 and 4 Mbps per subscriber.

Building energy management system (BEMS) (Chapter 2) BEMS is a class of technologies, processes, products, and services whose purpose is to address EE in commercial buildings. BEMS includes energy management technologies, submetered systems and software to provide building managers with energy usage data and feedback, but also lower tech solutions such as as spray foam and double-paned windows.

Business process improvement (BPI)   (Chapter 2) BPI is a business optimization mechanism whose goal is to reduce variation and/or waste in processes, in order to achieve a desired outcome with better resource utilization. As such, BPI focuses on individual processes and looks for efficiencies.

Carbon dioxide (CO_2)   A principal byproduct of fossil fuel combustion, the excess production of CO_2 is a key component of greenhouse gas emission, which are widely acknowledged as a principal cause of climate change and therefore a target for policy changes to address climate change.

Carrier   Another term for a wireless network operator, examples include ATT, Sprint, T-Mobile, and Verizon.

Cathode ray tube (CRT) screens   (Chapter 1) Video display units that use an electron gun inside a vacuum tube to shoot electrons at a fluorescent screen to produce images. The most common examples of CRT screens are television sets and computer monitors. CRT screens are largely being replaced by video screens using liquid-crystal display (LCD) and light-emitting diode (LED) technologies today.

Center for the Commercialization of Electric Technologies (CCET)   (Chapter 6) CCET is an Austin, Texas–based 501(c)(6) nonprofit whose goal is to collaboration between the electric industry, technology companies and universities in order to enhance the safety, reliability, security, and efficiency of the Texas electric transmission and distribution system. It accomplishes this forward-looking vision through research, development, and commercialization of emerging technologies with an emphasis on collaboration and process. Launched in 2005, CCET counts as members 21 Texas electric and high-tech companies and five universities. CCET is the recipient of a unique DOE FOA 36 smart grid Demonstration Grant, which ties together West Texas wind farms, synchrophasors, and a neighborhood in North Houston equipped as a VPP.

Code division multiple access (CDMA)   A method pioneered by Qualcomm for transmitting signals over wireless networks. In CDMA, many radios transmit and receive on the network at the same time, making it very efficient. In the United States, Sprint and Verizon use CDMA technology.

Combined heat and power (CHP)   (Chapter 2) Previously referred to as cogeneration, CHP is the simultaneous production of power and the capture of exhaust heat to provide thermal energy, to be used, for instance, to heat water for a district energy plant. With such dual energy production, CHP is one of the most efficient fossil fuel power sources. CHP is prevalent in industrial applications, but also can be used for residential (microCHP) and commercial applications (miniCHP).

Common indexing protocol (CIP)   An Internet protocol for finding information about resources on networks, CIP is the modern evolution of Whois++, its predecessor. CIP allows servers to know the contents of other servers through the exchange of index information, providing versatility and efficiency in information management. As an indexing protocol to define methods of creating and exchanging index information among indexing servers, CIP distributes searches across several instances of a single type of search engine to create a global directory and can tie individual databases into a distributed data warehouse.

Community energy storage (CES) The application of ES technology that is focused on supporting a small number of residential or commercial load centers, akin to a neighborhood ES facility, considered by some to be the sweet spot for immediate application based on current technology maturity, capabilities, and price points. AEP has been a principal advocate of CES applications.

Community solar A midsized solar PV system (100–500 KW) that produces energy for use by a group of individuals (community). *See also* solar garden.

Compact fluorescent lighting (CFL) A technology substitute for incandescent light bulbs that offers far more favorable energy ratings, CFL bulbs have become common. Replacement of incandescent light bulbs, which direct far more energy to heat production than light, is considered to be the easiest, cheapest, and most readily available energy efficiency step a household can take. Many utilities have CFL replacement programs and rebates.

Critical infrastructure protection (CIP), NERC CIP (Chapter 2). The CIP program coordinates NERC's efforts to improve physical and cybersecurity for the bulk power system of North America with regard to reliability and includes standards development, compliance enforcement, assessments of risk and preparedness, disseminating critical information via alerts to industry, and raising awareness of key issues. Additionally, the program is home to the Electricity Sector Information Sharing and Analysis Center (ES-ISAC), and it monitors the bulk power system to develop real-time situation awareness leadership and coordination services to the electric industry.

Critical peak pricing (CPP) An alternative pricing mechanism for electric utilities that focuses on the critical peak times—hours in the day and days in the season when the cost to produce energy is the highest, or when the stress on the system is the greatest. CPP is a strategy to shift consumption away from these peaks to drive down the operational costs of the utility.

Current state architectural framework (CSAF) A planning mechanism to capture the status quo of the utility architecture as a baseline for comparison with a planned architectural framework. The goal of this planning is to provide a gap analysis to define the steps for a smart grid transition.

Demand response (DR) DR systems consist of a remote control unit connected to a wireless network, used to automate load curtailment as an alternative to dispatching additional supply resources. This component of a smart grid project involves provisioning telecommunications connectivity, generally wireless or BPL, between the DR device, generally an HEMS, smart thermostat or a load controller, and the energy control center and hardware and applications that enable interoperability and automated response using separate or shared databases.

Demand response management system (DRMS) A system that aggregates one or more DR technologies to provide a single management view for a utility, enabling curtailed load to act as a dispatchable resource.

Department of Energy (DOE) (Chapter 1) With roots in the Atomic Energy Commission, formed after World War II to regulate the development of nuclear energy, the modern incarnation of the DOE was launched by the United States in October 1977, when the new federal department was created out of the Federal Energy Administration, the Energy Research and Development Administration, the Federal Power Commission, and parts and programs of several other agencies. Today, DOE is the focal point for administration of federal energy policy. Within the DOE, smart grid policy implementation is most influenced by activity in the Energy and Environment division and the National Energy Laboratories.

DOE FOA 36   (Chapter 5) The 2009 ARRA (stimulus bill) included funding to stimulate smart grid demonstrations, to be administered by the DOE in Funding Opportunity Announcement (FOA) 36 (smart grid demonstration matching grants), totaling over $435 million for regional demonstrations and $185 million for ES projects. The grant awards were announced the week of Thanksgiving in November 2009, with about 12 significant regional demonstration grant awards ranging from highs of $88 million (Pacific Northwest), $75 million (AEP Columbus), and $60 million (Los Angeles Department of Water and Power) to the smaller projects discussed in [Chapters 4 and 6—about $12 million (CCET) and $10 million (Pecan Street Project].

DOE FOA 58   (Chapter 6) The 2009 ARRA (stimulus bill) included funding to stimulate smart grid investments, to be administered by the DOE in Funding Opportunity Announcement (FOA) 58 (smart grid investment matching grants), totaling about $3.4 billion. The grant awards were announced in October 2009, with about 100 divided into two groups, 25 large projects (over $20 million, receiving approximately 80% of funding) and 75 small projects ($20 million or less, receiving approximately 20% of funding). The awards were in five categories.

Digital transition   This term describes the process all organizations must go through in the twenty-first century, as they leverage new technologies that provide new options for applications, equipment, processes, and networks that make them more effective. A good way to understand this issue is to consider two technology approaches, analog versus digital. Analog technology was revolutionary in the twentieth century, when radio and television changed the landscape through exploitation of better understanding of how radio frequencies behaved. With the advent of the transistor and the integrated circuit, however, a digital alternative was born and it matured in the second half of the twentieth century. As this digital progress was employed with the Internet at the turn of the century, the potential of the transformation became apparent, and companies began to leverage the new tools to be more competitive. Thus, the "analog" approach reflects a twentieth-century mindset that still relies upon paper-based data, labor inputs, and manual processes. In contrast, a "digital" approach demonstrates a twenty-first century perspective that takes advantage of low-cost, high-power digital computers and storage devices, VOIP communication devices, and broadband networks to transform the potential of organizations. Undergoing a digital transformation is a complex task that starts with a paradigm shift regarding the nature of the job, and a rewriting of the processes used to accomplish business objectives. Because digital technology evolves rapidly, a digital transformation is more of an ongoing process than it is an event with a beginning and an end.

Dispatchable   Utility operators dispatch power from central generation resources to keep the grid voltage levels in balance. A challenge to renewable energy resources like wind and solar energy is that the power they produce is intermittent, due to the nature of wind and sunlight. Consequently, the power produced by renewable energy resources is not considered "dispatchable" by grid operators and must be backed up by gas or coal generation resources in the event they suddenly become unavailable. In smart grid terms on the distribution grid, DER resources will be deemed dispatchable generation resources if they are equipped with communication devices and ES, so that grid operators can dispatch them similar to the way they dispatch central power today. DR resources may also be considered as dispatchable generation resources if they consistently curtail on demand with a record of that curtailment.

Distributed control system (DCS) (Chapter 2) DCS is used to connect the central power plants of a utility with its control center for generation dispatch. This component of a smart grid project involves provisioning high-speed connectivity, generally fiber optics, between the plants and the energy control center and applications that enable interoperability and automated response using separate or shared databases.

Distributed energy resources (DER) (Chapter 1) A new class of technology-driven edge power devices and systems that includes DG—most especially solar PV systems; EVs and charging stations, and ES. In the Pecan Street Project, this term was expanded to include DR, characterized by HEMSs, BEMSs, and IHDs.

Distributed energy resource optimization engine (DEROE) DEROE has a subset focus in functionality to that of the SGOE. The DEROE can aggregate, monitor, and manage systems, such as HEMS, BEMS, DRMS, DEMS, and ILM. The DEROE is an aggregation solution to manage a subset of systems on the edge that the SGOE could ultimately monitor and manage. However, the DEROE could remain as a stand-alone system managing multiple microgrids.

Distributed energy management system (DEMS), distributed energy resource management system (DERMS) DEMS and DERMS, like their counterpart DRMS, are management systems used by utilities to gain dispatchable control over multiple DG systems.

Distributed generation (DG) (Chapter 1) A class of energy devices and systems characterized by their location at the edges of the distribution grid, near to the load that they serve. DG includes such clean energy technology as rooftop solar PV systems, micro gas turbines using CHP, and micro wind turbine systems, but also more traditional portable gas and diesel generators.

Distributed independent power producer (Chapter 7) A PEB that sells its excess power to the utility or into the market.

Distributed network protocol, version 3 DNP3 DNP3 is a Layer 2 networking protocol widely used in SCADA applications, enabling communications for SCADA master controllers, RTUs, and IEDs.

Distributed solar photovoltaic (Chapter 6) A small energy production system that uses solar panels to produce energy at a site where the energy is consumed, stored, or sold over the grid, typically sized under 100 KW.

Distribution automation (DA) A class of applications and systems that automate the monitoring, management, and control of the distribution grid, from substations to distribution feeders. DA systems include GISs, OMSs, and DMSs. DA is the key to a new area of utility benefits known as gird optimization, where automated processes lower reserve thresholds, eliminate line losses, and fine-tune the grid.

Distribution management system (DMS) (Chapter 1) A DMS automates many distribution functions and proactively responds to outage information while integrating all the features of a traditional OMS and enabling dynamic GIS evolution.

Distribution service platform provider (DSP) The N.Y. PSC introduced this concept in its 2014 regulatory proceeding titled REV ("Reforming the Energy Vision"), suggesting the need for a new approach to managing an ever more complex distribution environment. A DSP, which could be a utility or a nonutility, has the role to provide a base platform on which will reside innovative applications and platforms that integrate DER with traditional grid resources to provide new value-added services, enhanced system reliability, and more environmentally sound operations. The goal of the DSP is to bring system coordination down to the distribution level, provide local and regional management and oversight, and ensure that new technologies are introduced deliberately to support a managed transition.

Distribution system operator (DSO)* (Chapter 5) As distribution utilities introduce DER onto their grids, their responsibilities and operational functionality will move closer to that of an ISO. Those utilities may then choose to become a DSO, a term used in Pecan Street Project discussions to describe a new role for distribution utilities that are implementing advanced smart grids. The utility that integrates DER activities will need to have systems and processes to manage market functions and the flow of energy across its grid, like those at an ISO.

Edison Electric Institute (EEI) (Chapter 6) As the association representing IOUs, EEI provides smart grid online resources, reports, workshops, and focus at its conferences, roundtables and seminars. For example, at its roundtable in October 2010, EEI presented a smart grid scenario project update, showcasing results of its two workshops that year on smart grid, held in Washington, D.C., and Los Angeles.

Electric Power Research Institute (EPRI) (Chapter 1) EPRI is the nonprofit R&D arm of the electric utility industry, focused on improvements in reliability, efficiency, health, safety, and the environment. EPRI members represent over 90% of U.S. generation and over 40 countries worldwide. EPRI offices are found in Palo Alto, CA; Charlotte, NC; Knoxville, TN; and Lenox, MA. EPRI was an early proponent of smart grid with its Intelligrid concept nearly 10 years ago.

Electric Reliability Council of Texas (ERCOT) (Chapter 4) ERCOT is a nongovernmental entity charged with the management of electric power in a territory largely coterminous with the State of Texas, about 75 percent of the Texas land area, with portions of northwest, northeast, and southeast Texas outside ERCOT boundaries. ERCOT manages electric power service to 22 million Texas customers, representing 85 percent of the state's electric load. As regional ISO, ERCOT schedules power on an electric grid that connects 40,000 miles of transmission lines and more than 550 generation units. ERCOT also manages financial settlement for the competitive wholesale bulk-power market and administers customer switching for 6.5 million Texans in competitive choice areas.

Electric vehicle (EV) (Chapter 2) EV includes the electric or plug-in hybrid electric vehicles, electric charging stations, and respective supporting networks. Charging stations are likely to be deployed at residences and businesses, as well as at public locations including charging stations available to the public at the curbside and in parking garages and parking lots.

Electric vehicle support equipment (EVSE) EVSE, the charging system complement to the electric vehicles in the EV category, consists of all equipment needed to establish a charging function for the EV.

Energy consumption versus energy demand A distribution grid is designed and built to service a projected system peak load—the time during the year of maximum energy consumption, when all generation units are operating at full capacity to meet demand and the distribution grid is at peak capacity as well. Energy consumption, measured in kilowatt-hours, is the cumulative amount of energy represented by the area under the load curve over a specific amount of time. Energy demand, measured in kilowatts, is the highest point of the load curve over a specific amount of time. Residential consumers typically are charged only for consumption, while commercial and industrial consumers pay for both consumption and demand. This is because rates are designed to recover both capital and operational costs. Electric utilities invest capital for generation, transmission, and distribution assets—capacity to meet peak demand, but also operational expenses such as fuel costs to produce energy.

Energy efficiency (EE)   (Chapter 2) EE is a class of smart grid that is focused on the demand side of the equation, on energy consumption. The built infrastructure consumes approximately 70 percent of the energy produced, so focusing on consumption becomes a way of operating a more efficient system. EE may be divided into passive strategies, principally focused on sealing the building envelope as a more effective container of heat or cold (e.g., insulation and radiant barrier) and on upgrading energy appliances to the most efficient models and technologies (replacing incandescent light bulbs with CFLs and LEDs, switching HVAC to a higher SEER rating). EE may be low-tech (turning off light bulbs when leaving the room, sealing cracks around doors and windows) or high-tech (spray foam insulation, programmable HEMS). As a component of smart grid, EE is often left out as it is considered low-tech and not connected to the grid. However, a focused strategy on EE may be seen as an effective complement to a smart grid program, in as much as a system should be designed to serve a targeted system load, and making the load efficient after the smart grid is designed may result in an over-designed system.

Energy management system (EMS)   (Chapter 2) EMS is a general category of computer-aided tools that electric utility operators use to monitor, control and optimize system performance at the generator and transmission levels. Combined with SCADA (EMS/SCADA), EMS refers to the optimization component of advanced applications for generation control and scheduling.

Energy municipal utility district (eMUD).*   (Chapter 7) A MUD is a strategy to provide public utility infrastructure where city systems are not available (see MUD). Over time, some MUDs have expanded their purview to include other services such as garbage collection. Given the advances to energy technologies as seen in such categories as DG, community ES, and aggregated DR, we project the creation of an eMUD to provide a local community with new options for energy.

Energy storage (ES)   (Chapter 2) ES is an emerging smart grid category, perhaps the least mature of the major components. ES may be fixed or mobile, as in the case of EVs. ES comes in different sizes, ranging from personal ES serving one home or building, to community ES serving a neighborhood or group of homes or buildings, to utility-scale ES, as a component of the distribution grid used for energy balancing, ancillary services, renewable energy integration, or arbitrage. Principal types of energy storage include (1) mechanical—e.g., hydroelectric (pumped hydro), compressed air ES (CAES), and flywheels; (2) electrochemical—e.g., lead acid batteries, advanced lead acid batteries, sodium sulfur and flow batteries, and fuel cells; (3) electrical—capacitors, supercapacitors, and ultracapacitors; (4) thermal—e.g., ice and molten salt; and (5) chemical, biological, or the like. ES holds great potential as a disruptor to the current electricity paradigm, which has generally been a just-in-time system, designed and operated in just-in-time fashion, without the capacity for managing peaks with the benefit of stored energy.

Energy Storage Association (ESA)   (Chapter 6) The ESA, an international trade association promoting the development and commercialization of ES technologies, has been hard at its task for nearly two decades. Highly complex and varied, ES technologies face a significant challenge in the widespread misperception that energy cannot be stored economically. Recent technological progress has chipped away at that attitude, as economically viable utility-scale storage solutions become increasingly available. ES took a big leap forward with the enactment of an energy storage bill, AB 2514 by the California legislature, in October 2010.

Environmental Defense Fund (EDF)   (Chapter 6) Formed in 1967, EDF has long been an outspoken advocate for the environment. Devoting significant resources and leadership to the Pecan Street Project in 2009, EDF has developed a strategic interest in electric utility smart grid projects and is promoting clean energy best practices to the sector in alignment with the organization's environmental goals.

Ethernet   A frame-based computer networking technology for local area networks (LANs) that defines wiring and signaling for the physical layer and frame formats and protocols for the media access control (MAC)/data link layer of the OSI model. Ethernet is mostly standardized as IEEE 802. It has become the most widespread LAN technology in use since networking became prevalent in the 1990s and has largely replaced all other LAN standards.

Fault detection, isolation and restoration (FDIR), predictive FDIR   (Chapter 2) FDIR is a function of the distribution grid for outage management and restoration and includes three components: (1) fault detection (a monitoring event); (2) fault isolation (a control event); and (3) restoration to normal grid conditions (a management event). These steps will one day be taken automatically by digital devices attached to edge devices, according to preprogrammed algorithms, to provide predictive FDIR, a new capability to address grid conditions, this time by using prediction to improve recovery when things go wrong.

Federal Energy Regulatory Commission (FERC)   (Chapter 2) The FERC is the independent regulatory agency with purview over the interstate transmission of electricity, natural gas, and oil. From an electric utility industry perspective, FERC regulates transmission and interstate wholesale power markets, oversees the M&A and corporate activity of electricity companies, reviews transmission project siting applications in some instances, licenses and inspects hydroelectric projects, protects the reliability of high-voltage interstate transmission system via mandatory reliability standards, monitors and investigates energy markets, and administers accounting and financial reporting regulations and the conduct of regulated companies. FERC chairman Wellinghoff has taken a special interest in EV and DR subjects over the past year.

First-generation smart grid (smart grid 1.0 or SG1)*   (Chapter 1) A smart grid project that begins with an application, such as AMI, to address a need in one of the utility departmental silos, without a prior, deliberate effort to craft a smart grid architecture framework and design.

Geospatial information system (GIS)   (Chapter 1) A system that captures, stores, analyzes, manages, and presents data that are correlated to specific geospatial or geographic location(s). GIS combines cartography, statistical analysis, and database technology and is a widely used tool in electric utilities to manage the distribution grid assets and ensure reliability. GIS is a component of a DA solution.

Government Accountability Office (GAO)   (Chapter 2) The U.S. GAO is an independent, nonpartisan agency that serves the U.S. congress as a watchdog, investigating how the federal government spends taxpayer dollars. The GAO is headed by the comptroller general of the United States, which is appointed by the president and confirmed by the U.S. Senate. U.S. congressional committees and subcommittees direct the GAO to undertake research, but research projects also stem from public laws and committee reports, or by independent direction of the comptroller general. GAO audits agency operations, investigates allegations of law-breaking and impropriety, reports on the efficacy of government programs and policies, performs policy analysis and outlines opinions, and issues legal decisions and opinions.

GridWise Alliance (GWA)   (Chapter 6) The GWA coordinates smart grid organizations by facilitating activities between stakeholder groups and by developing a sound foundation of educational and policy materials, acting as the go-to industry representative for government policymakers and the press when it comes to all issues associated with smart grid. The GWA accomplishes its mission through work groups (e.g., the Implementation Work Group focuses on smart grid case studies and value streams for stakeholders).

GridWise Architecture Council (GWAC)   (Chapter 6) Neither a design team nor a standards making body, the GWAC is a group of industry leaders, formed at the direction of the DOE, to help shape the architecture of the emerging smart grid. Its principal focus is to provide guidelines for industry interaction and for interoperability between technologies and systems. The GWAC seeks to identify areas for standardization that will stimulate interoperation between the components of the smart grid.

Heating, ventilation, air conditioning (HVAC) systems   HVAC systems manage temperature and humidity environment for buildings and homes, so the term includes heaters, boilers, air conditioners, chillers, and refrigeration units. For smart grid purposes, HVAC systems matter because they are often the largest single energy-consuming appliance inside the building or home. For DR, HEMS units interface with the HVAC, providing feedback on energy consumption, but more importantly, direct control to cycle the HVAC on and off to reduce energy consumption during peak periods.

Home energy rating system   (Chapter 7) HERS is the industry standard index by which a home's EE is measured. It's also the nationally recognized system for inspecting and calculating a home's energy performance.

High-penetration PV (HPPV)   (Chapter 7) HPPV describes efforts to load up a single distribution feeder with higher concentrations of PV facilities than current standards allow. As a rule of thumb, a single distribution feeder can only handle about a 20 percent penetration of PV, beyond which the potential for intolerable risk and instability to grid operations sets a boundary. The intermittency of production and the potential for excess power to reverse the power flow on distribution feeder lines pose a threat to upstream equipment. HPPV anticipates new technologies and processes that will enable the distribution grid to safely accommodate large concentrations of edge power.

Home energy management system (HEMS)   HEMS is an acronym for a new class of consumer devices that combine equipment and software to provide consumers with a feedback mechanism to better monitor and manage their in-home energy consumption. Closely related to home area networks (HANs), HEMS may include a connection to the smart meter, drawing revenue data from the meter via ZigBee connectivity. HEMS will enable DR services to the degree that they automate or facilitate consumer decisions to curtail power during peak periods at the request of the utility.

Independent system operator (ISO)   (Chapter 4) The FERC directs the formation of ISOs for individual states, but also sometimes regions, as a coordination and monitoring mechanism to provide oversight and direction to regional electric grids for reliability purposes. An ISO may also have the purview over market operations of the wholesale power market in a region, overseeing power dispatch and grid balancing, but also providing a market clearing function.

Information and communication technology (ICT)   (Chapter 4) A broader term than IT, ITC includes telecommunications technology as well.

Information technology (IT)   (Chapter 1) Perhaps one of the broadest terms in use today, IT refers to anything that provides data, information, or knowledge in any visual format using any multimedia distribution mechanism. IT is the foundation of our modern economy, ranging from computers and hardware to software to communications.

Information Technology Infrastructure Library (ITIL)   (Chapter 4). ITIL is a set of concepts and practices for IT professionals to manage and run their operations better, providing detailed descriptions of IT practices and comprehensive checklists, tasks and procedures that can be tailored to meet the specific needs of an organization.

In-home display (IHD)   (Chapter 2) This term refers to any type of digital information display inside a residence. In the context of this book, IHDs are used in HEMS to provide feedback on energy consumption to the energy consumer.

Institute of Electrical and Electronics Engineers (IEEE)   (Chapter 2) Formed at the dawn of the electric industry, the IEEE began in the spring of 1884 in New York as the American Institute of Electrical Engineers (AIEE), drawing its membership from the telegraphy, electricity, and telephone industries. The AIEE led the growth of the electrical engineering profession, and when wireless became big 100 years ago, the Institute of Radio Engineers modeled itself on the AIEE. By 1963, both industries had grown, and as their purviews overlapped, they decided to merge to create the IEEE, which had 150,000 members at that time, all but 10,000 in the United States. The IEEE has continued to grow and address issues in the electronics industries—by the early twenty-first century, it served its members with 38 societies, 130 journals, and more than 300 conferences annually, supporting over 900 active standards. By 2010, IEE could truly claim to represent the world's engineering community as the world's largest technical professional association, with nearly 400,000 members in 160 countries.

Intelligent load management (ILM)   ILM is a new term used to describe the union of DRMS and DEMS, enabling the control of edge resources under one management system, both loads and resources. See also DER optimization engine.

Internal combustion engine (ICE)   ICE, used to describe engines that run on gasoline, diesel, and natural gas, is a term that has grown in use in discussions about EVs.

International Electrotechnical Commission (IEC) (Chapter 2) The IEC is the global body that establishes and administers international standards for devices and systems that contain electronics and use or produce electricity through the IEC International Standards and Conformity Assessment Systems to ensure that those devices and systems perform, fit, and work safely together. Founded in 1906, the IEC is the world's leading organization for the preparation and publication of international standards for all electrical, electronic, and related technologies.

International Standards Organization (ISO) (Chapter 4). ISO is an NGO and the world's largest developer and publisher of international standards. ISO is actually a consensus-oriented network of national standards institutes from around the world (160 countries, one member per country), with a central secretariat in Geneva, Switzerland, as coordinator. The ISO 9000 family of standards, for instance, provides a template for organizations for quality management, designed to help ensure that organizations meet the needs of internal and external stakeholders.

Internet of things (IoT) (Chapter 7) IoT is the interconnection of uniquely identifiable embedded computing devices within the existing Internet infrastructure. Typically, IoT is expected to offer advanced connectivity of devices, systems, and services that goes beyond machine-to-machine (M2M) communications and covers a variety of protocols, domains, and applications. The interconnection of these embedded devices is expected to usher in automation in nearly all sectors of the economy and society, while also furthering advanced applications like smart grids. Current utility examples include smart thermostat systems and home appliances for remote monitoring via the Internet.

Internet protocol (IP) (Chapter 1) The protocol within transmission control protocol/Internet protocol (TCP/IP) that is used to send data between computers over the internet. More specifically, this protocol governs the routing of data messages, which are transmitted in smaller components called packets. Devices that use IP are speaking a common language, using the Internet as their communication network.

Investor-owned utility (IOU) One of three principal classes of electric utilities in the United States, IOUs are owned by private investors and are typically larger than the other two types of utilities, municipally owned utilities (MOUs) and electric cooperatives (co-ops). IOUs are regulated at the state and federal level. The EEI is the industry body most closely associated with IOUs.

Islanding (Chapter 7) The voluntary or involuntary off-grid functioning of a premise, community, or local area that has the capability to provide power for itself (also resource islanding).

Key performance indicator (KPI) (Chapter 4). As the saying goes, "You can't manage what you can't measure," and KPIs are the way organizations measure what they manage. KPIs are units that measure the progress of entire organizations or divisions within the organization toward a goal, either tactical or strategic, inside an organization. As a performance metric, KPIs have value in measuring the attainment of objectives in a quantifiable way and provide the organization a way to improve on past performance in an objective manner.

Killer app A term devised to explain what happens when a computer application is so popular that it leads consumers to adopt a new technology in droves. For instance, e-mail and the Netscape Web browser are described as killer apps with regard to the early days of dial-up Internet access. In more current smart grid terms, the EV has been described by some as a killer app for smart grid, given that it will lead consumers to become more interested and invested in electricity consumption and production, and it will lead utilities to invest in grid upgrades to prevent damage to expensive distribution equipment.

Kilowatt (kW), kilowatt-hour (kWh)   (Chapter 1) Kilowatt is a term to define energy generation capacity equal to 1,000 watts, whereas kilowatt-hour is a common unit of measurement for electric power consumption equal to 1,000 watts in one hour or 3.6 megajoules. The kilowatt is used as a billing unit by electric utilities for energy demand, whereas the kilowatt-hour is used to measure and bill the amount of energy consumed.

Large Public Power Council (LPPC)   A subset of the APPA, the LPPC includes the largest of the publicly owned electric utilities.

Leadership in Energy and Environmental Design (LEED)   (Chapter 3). LEED is a classification scheme developed by the USGBC that provides guidelines and independent certification for builders that focus on unique design to promote not just efficient use of energy, but also more efficient use of water, lower CO_2 emissions, and improved internal comfort for the building occupants. LEED rates buildings and building designs in Silver, Gold, and Platinum levels, indicating the degree of adherences to LEED guidelines.

Levelized cost of energy (LCOE)   LCOE, measured in kilowatt-hours or megawatt-hours, is a way of comparing energy from different generation sources, using such measurements as initial capital investment, ROI, operating costs, fuel cost, and maintenance costs.

Light-emitting diode (LED)   LEDs, a light from semiconductor chips, is a technology that has advanced over nearly 50 years to become an alternative to incandescent light bulbs and compact fluorescent light bulbs. Advantages of LEDs are numerous: lower energy consumption, longer life, smaller size, faster switching, and greater durability and reliability. The principal advantage of LEDs is their low operating costs, but that is offset by their still relatively high upfront costs. LEDs are increasingly in use in small form factors such as flashlights and decorative holiday light strings.

Local area network (LAN)   LAN is a wired or wireless network connecting two or more computers or other devices over a short distance, such as within an office or a home. Wi-Fi is a wireless LAN technology.

Long-playing vinyl record (LP)   Offered as something of a tongue-in-cheek glossary item, the long-playing vinyl record or LP is a term widely understood, but perhaps not as much for the under-30 set of readers. An LP is offered as a great example of a successful analog recording technology at 33 1/3 RPM, the record album of popular music, which has been almost entirely replaced by digital substitutes, first by digital compact discs (CDs), then more recently by software such as MP3 and iTunes. While analog-to-digital transition in this instance dramatically lowered the cost of recorded music, more accurately captured sounds, and eased distribution and consumption of recorded music, it left behind a tonal quality that is missed by audiophiles, so the LP persists as a niche technology, and an object lesson that not all digital transitions will be complete.

Long-term evolution (LTE)   A term used to describe the upcoming generation of mobile wireless telecommunications. See 4G.

Maslow's Hierarchy of Needs   (Chapter 1) Often shortened to "Maslow's Pyramid," this hierarchy is an attempt to capture and prioritize the fundamental needs for living a human life. Starting with physiological, the pyramid proceeds to add mounting layers of needs including such categories as safety, love/belonging, and esteem, culminating in self-actualization. According to Maslow's theory, until the basic needs are met at the bottom of the pyramid, the higher needs are not or cannot be pursued. In other words, a person who is starving doesn't have time or attention for issues of morality or creativity. We propose that electricity be added at the base. A person lacking electricity becomes most concerned with a return to the previous condition of having electricity.

Megawatt (MW), megawatt-hour (MWh) A megawatt is a measure of energy capacity, a million watts, and is generally the measurement used to describe the capacity of large generators. In contrast, a megawatt-hour is a power measurement, multiplying the capacity over a period of one hour. A 100-MW generator produces 100 MWh in one hour. Collectively, 1,000 homes consuming 1,000 kWh each month consume 12 MWh over one year.

Meme A relatively modern term, a meme refers to ideas or beliefs transferred from a person or group to another, mimicking the process in nature for transferring genetic material from one generation to the next with genes and borrows from the word "mime" as in "to imitate." The term was first used in the 1970s to explain how evolutionary principles such as natural selection could apply to ideas and cultural themes as well.

Metcalfe's Law (Chapter 1). An observation first made in the early 1980s and attributed to Robert Metcalfe, a co-inventor of Ethernet, used to describe the increase in the value of a telecommunications network as being proportional to the square of the number of connected users of the system (n^2). Metcalfe's Law explains why networks such as the Internet have increasing returns—the larger the network, the greater the advantage to each participant on the network. Each new participant brings ever more value to the overall system. Thus, according to Metcalfe's Law, a network with only 10 users connected to it would have a theoretical value of 100, whereas a network with 100 people connected to it would have a theoretical value of 10,000—10 times more participants result in 100 times more value for all participants. Prior to the Internet, a principal example of Metcalfe's Law in the 1980s had to do with the value of fax machines—the more that were connected, the more valuable it became to own one. Now widely applied to the Internet, and also used in analysis of economics and business, Metcalfe's Law has become shorthand for a more general observation of how valuable a network becomes as it begins to add users and gain acceptance. Some have challenged the actual mathematics of the law, but suffice it to say that a network grows rapidly in value as nodes are added and this concept supports the expansion of the advanced smart grid.

Meter data management (MDM), meter data management system (MDMS) (Chapter 3) MDM, is a relatively new term, but a very important one for the smart grid. The amount of data coming from digital smart meters will dramatically increase as more and more AMI systems are deployed, making management of that data a new critical skill for utilities. The raw data that comes from the smart meters is stored in head end servers, processed (see VEE), and then used for digital billing and other analytical purposes. MDMS is the term for the system that performs the MDM function. MDM is a vital component of the value production of smart grid, given that data will feed the variety of applications inside a utility to drive processes and deliver desired outcomes.

Metropolitan statistical area (MSA)   As defined by the U.S. Office of Management and Budget: "a county or group of contiguous counties that contain (1) at least one city of 50,000 inhabitants or more (or 'twin cities' with a combined population of at least 50,000), or (2) an urbanized area of at least 50,000 inhabitants and a total MSA population of at least 100,000 (75,000 in New England)." The contiguous counties are included in an MSA if they are essentially metropolitan in character and are socially and economically integrated with the central city.

Microgrid   A microgrid is a modern, small-scale version of a utility-size power grid. A microgrid is used to achieve certain goals, such as more reliability, carbon emission reduction, diversification of energy sources, and cost reduction established by the single customer or community of customers being served. A microgrid can serve a variety of users such as one building, an office campus, a neighborhood, a new small village, or a small independent section of a city. Like the bulk power grid, a microgrid generates, distributes, and regulates the flow of electricity for its consumers, but does so locally. A microgrid is an ideal way to integrate renewable resources on the community level and allow for customer participation in the electricity enterprise.

Mobile virtual network operator (MVNO)   A company that provides operator services, but does not have a physical network, an MVNO purchases capacity from a facilities-based network operator to provide mobile services to customers. By licensing and reselling others' network and device infrastructure and services, MVNOs acquire the systems capability necessary to provide services and roaming.

Moore's Law   (Chapter 1). Like Metcalfe's Law above, Moore's Law is based on an observation regarding increasing value, this time with regard to the trend line of miniaturization of transistors and how many can be placed on an integrated circuit. Moore's Law observes that every two years, the number of transistors on an integrated circuit doubles. Expanded beyond transistors, this observation of steady, exponential value growth applies equally to other areas of the digital economy, including processor speeds, memory capacity, and the number of pixels in digital cameras. With regard to advanced smart grids, we project that Moore's Law will apply to the addition of sensors on the grid, driving exponential value creation in energy production, distribution, and consumption.

Multidwelling unit (MDU)   (Chapter 4) MDU is an industry term that is well described by its full name. Examples of MDUs include apartments, condominiums, dormitories, duplexes, and multiplexes. The significance for MDUs in smart grid discussions is (1) their limited ability to participate in DER utility programs (due to such factors as no solar rooftops or EV charging locations) and (2) their potential for submetering, among other capabilities.

Municipally owned utility (MOU)   Community-owned utilities, known in the industry as MOUs, are city departments operated by city employees with direct or indirect board oversight by city government (city councils or independent boards). Local, hometown decision-making is a critical element supporting MOUs, as well as ensuring access to a stable supply of electricity while protecting the environment. About two-thirds of these utilities do not generate their own electricity, but instead purchase wholesale power to distribute to their citizens/customers.

Municipal utility district (MUD)   (Chapter 7). A MUD is a special-purpose district created by governments to provide public utilities to local residents. MUDs are formed with a vote and then represented by a board of directors also created by local vote. MUDs are empowered in their charters with such rights as seizing land under eminent domain, levying taxes, and raising bonds.

National Association of Regulatory Utility Commissioners (NARUC) (Chapter 6) NARUC, a national nonprofit organization, represents the state commissions that regulate electricity, telecommunications, water, and transportation. NARUC holds semiannual meetings throughout the country where commissioners gather to accomplish the work of the association, attending committee meetings, sharing best practices and comparing notes, and meeting with utility, consumer, and vendor representatives to stay informed on industry perspectives. The NARUC-FERC Smart Grid Collaborative provides a forum for state and federal regulators to discuss a range of issues to help facilitate the transition to a smart grid, with a special focus on smart grid technologies. At the beginning of September 2010, NARUC announced the formation of a Smart Grid Working Group to be comprised of seven state commissioners representing the diversity of smart grid in the United States, to be cochaired by commissioners from New York and Michigan.

National Association of State Utility Consumer Advocates (NASUCA) (Chapter 6) For over 30 years, NASUCA has provided a forum for organizations that represent utility consumers in regulatory and court proceedings. Today, membership has grown to 44 consumer advocate organizations from 40 states and the District of Columbia. In 12 states, consumer advocacy is handled by state attorneys general, while in 29 others, consumer advocacy offices have that role, with directors appointed by governors. Traditionally, NASUCA member-organizations have had the role of challenging rate increases in adversarial rate cases. NASUCA filed comments in August 2010 to a DOE RFI on smart grid, focusing its argument on the ability of customers to opt in to advanced rates, the importance of cost-benefit justification of smart grid projects, and the need to educate the widely divergent array of customers on smart grid impacts and issues in tailored programs.

National Institute of Standards and Technology (NIST) NIST is an agency of the U.S. Department of Commerce. Founded in 1901, NIST is the nation's first federal physical science research laboratory. Title XIII of the Energy Independence and Security Act (EISA) of 2007 gave NIST primary responsibility to coordinate development of a smart grid framework with protocols and model standards for information management to achieve interoperability of smart grid devices and systems. NIST is also mandated to develop security standards for the smart grid.

National Renewable Energy Labs (NREL) (Chapter 6) Located in Boulder, Colorado, NREL is the primary R&D lab of all the DOE research labs for renewable energy and energy efficiency.

National Rural Electric Cooperative Association (NRECA) (Chapter 6) NRECA is a national service organization that represents the national interests of cooperative electric utilities and their consumers (the corollary to APPA and EEI). Founded in 1942, NRECA helped the emerging rural electric cooperatives, becoming an advocate for energy and operational issues as well as rural community and economic development. With more than 900 member cooperatives that serve 42 million people in 47 states, NRECA is made up mostly of member-owned cooperatives, with a handful of public power districts, as well as the generation and transmission (G&T) cooperatives formed by retail cooperatives for power supply and other classes of members.

National Rural Telecommunications Council (NRTC) (Chapter 6) Supporting both electricity and telecom rural cooperatives, NRTC provides technology and procurement support to facilitate telecom solutions.

Network layer Level 3 in the seven-layer OSI model, this layer is utilized by a segment of the ISP industry comprising the companies that buy access to the physical layer pipes of telephone companies. Those companies then build nationwide TCP/IP backbones and sell access on a wholesale basis to brand layer companies. UU-NET was the most successful early network layer company and sold access to brand layer companies such as Earthlink, AOL, and MSN. This group of companies forms the middle layers of the OSI model.

Network operations center (NOC) The physical command center of a telecommunications network, which monitors a large network 24 hours a day, a NOC is typically a room with monitors showing the real-time, detailed status of one or more networks and is the location of servers that run the network.

Network operator A company that provides wireless digital data services by assembling and managing the required equipment, sites, switches, lines, circuits, software, and other transmission apparatus used to provide telecommunications services.

North American Reliability Council (NERC) (Chapter 2) After the catastrophic blackouts in the northeastern United States in the late 1960s, the NERC was established as a nongovernment organization in 1968 to ensure the reliability of the bulk power system in North America. NERC develops and enforces reliability standards, leverages short-term and long-term forecasts, monitors the bulk power system, and provides education, training and certification for industry personnel. NERC announced a joint collaboration in 2011 with the DOE and NIST to develop and promote smart grid cybersecurity standards.

Open systems interconnection (OSI) (Chapter 1) The OSI model, often referred to simply as the "OSI stack," is a model developed by the ISO to explain the functionality of a communication system into layers that interoperate in a logical way, by providing unique servers up or down to another layer. The stack starts with Layer 1 and progresses to Layer 7.

Operational level agreement (OLA) (Chapter 4) The internal agreements within the different technology groups at Austin Energy, OLAs were modeled on the SLAs that vendors commonly use with their customers to define the service they provide. With written OLAs, Austin Energy's ITT department sought to gain organization commitment on its initiatives.

Original design manufacturer (ODM) ODMs are similar to OEMs, but instead of manufacturing products, they design them and have the manufacturing accomplished by outside firms.

Original equipment manufacturer (OEM) A term for any company in the computer industry that makes equipment for sale through a reseller to end users, including desktop computers, laptops, and networking equipment such as routers and Wi-Fi PCMCIA cards and access points. Examples of OEMs include Dell, HP, Sony, Apple, Proxim, Tropos, Linksys, Siemens, and Cisco.

Outage management system (OMS) (Chapter 1) An OMS is a software solution that integrates multiple systems inside an electric utility to help it manage and recover from an outage. To accomplish this difficult task, the OMS must integrate with the CIS, GIS, voice response systems, and AMR system or more recently the AMI system.

Peer-to-peer energy trading (P2PET)   (Chapter 7) P2PET will become a reality when the distributed generation (solar PV), ES, and DR are integrated on the advanced smart grid, and the utility adopts a distribution system operator (DSO) role. P2PET uses the distribution grid to wheel energy produced at the edge to an edge consumer. As such, P2PET is an accounting transaction between buyer and seller, with a distribution fee collected by the DSO and the need for a market clearing function, much as wholesale transactions are conducted today.

Photovoltaic (PV)   (Chapter 2) PV panels and film systems create electricity directly from sunlight and its interaction with the materials in the PV system. Silicon chips are mounted in solar cells on solar panels in rigid frames, while PV film has the active substance printed on a substrate (film) that can be applied to a variety of building materials (e.g., roofing tiles, wall panels, and windows) to create building integrated PV (BIPV).

Physical layer   In network communications, the physical layer is Level 1 in the seven-level OSI model of computer networking. The physical layer performs services requested by the data link layer. The physical layer is the segment of the ISP industry comprised of those companies that build and operate physical network infrastructure, such as regional Bell operating companies (RBOCs), long-distance carriers and cable companies, and those who sell wholesale access to network layer companies such as UUNET, which, in turn operate, nationwide TCP/IP networks. Physical layer companies include SBC, Verizon, and AT&T.

Plug-in hybrid electric vehicle (PHEV)   In contrast to a pure EV, which runs strictly on electricity, a PHEV has a gas tank and combustion engine as a complement to it electric batteries and power system. PHEVs have longer range but produce more greenhouse gas emissions. (*See* EV.)

Point-to-point (P2P), point-to-multipoint (P2MP)   These terms are used to describe wireless broadband technology that sends a signal directly from one point to another, or from one point to many, in contrast to a mesh network that sends signals indirectly across multiple nodes in series to find the destination. An example of P2P and P2MP technology is WiMAX.

Positive energy building (PEB)   (Chapter 7) A building designed to produce excess power so that it becomes a power plant as well as a residence or workplace, becoming a power plant itself (a negative HERS rating). A PEB is a building that is not only a residence or place of work, but also a source of income as a DIPP.

Positive energy mortgage (PEM)   (Chapter 7) PEMs can be expected as a financial innovation to accompany the rise of PEBs. Homeowners and businesses who design buildings as PEBs may choose to apply the surplus they receive from energy sales to accelerate the payoff of mortgages, rather than, or in addition to, selling the power as a DIPP. Ten-year PEMs would line up with ARMs as financial innovations in mortgage lending. The PEM would have accelerated payoff of the mortgage based on the income from the PEB. We may go even further to imagine PEMs with lower rates as well, because the added revenue makes PEM payback less risky.

Power line carrier (PLC)   (Chapter 1) An early technology to send a narrowband communication signal over a power line, PLC is still in use in rural areas for meter reading (e.g., DCSI and Hunt) and in local areas for communication (e.g., HomePlug).

Profit and loss statement (P&L)   A synonym for income statement, a P&L is an accounting tool to show investors and managers a periodic reporting on income and expenses, to track financial performance, including revenue, expenses, and net income. P&Ls can apply to companies or individual departments.

Prosumer   (Chapter 3). Futurologist Alvin Toffler coined the term *prosumer* in 1980. Loosely, Toffler's "proactive consumers" or prosumers were common consumers who were predicted to each become active to help personally improve or design the goods and services of the marketplace, transforming it and their roles as consumers. By far the most common usage of the term describes the consumers, enthusiasts who buy products (almost always technical) that fall between professional and consumer grade standards in quality, complexity, or functionality. "Producing consumer" prosumers create goods for their own use and also possibly to sell. Hence for the energy industry, a consumer that produces energy on his or her rooftop using solar panels and sells the excess back to the utility is an energy prosumer. Energy prosumers are highly knowledgeable, demanding consumers and are interested in producing their own clean energy to meet their social, environmental, and financial goals.

Public Service Commission (PSC), Public Utility Commission (PUC)   (Chapter 6. PSC and PUC are acronyms for state regulatory commissions. IOUs are regulated at the state level on a variety of levels, including ratemaking, cost recovery, consumer interaction, and siting decisions. A key element in smart grid cases that has arisen in 2010 is the value proposition for smart grid, specifically if the utility intends to recover costs in a rate proceeding.

Public Utility Holding Company Act (PUHCA)   (Chapter 7) Also known as the "35 Act," this federal legislation passed in 1935 was the keynote of FDR's financial reforms to address the financial excesses of the Great Depression. FDR's attorney general took aim at Samuel Insull and his holding company structure, a financial innovation that unlocked capital financing to build out the electric grid infrastructure. However, holding companies were controlled by a relatively small number of individuals who owned and managed multiple geographic electric utilities. Such concentrated ownership was called a trust, and PUHCA was the primary example of trust busting in the FDR administration. PUHCA would become a key driver of today's U.S. state regulatory system.

Rational unified process (RUP)   (Chapter 4) RUP is an iterative software development process created by Rational Software, a division of IBM. RUP is an adaptable process, where teams select the elements that fit their project.

Real-time pricing (RTP)   (Chapter 7) Rates based on TOU, RTP, and CPP help a utility differentiate its commodity kilowatt-hours by price in order to motivate consumers to shift their consumption to off-peak hours. RTP differs by having pricing closely track the cost of energy throughout a 24-hour period versus in three or four major periods (TOU), or focusing on the most extreme peaks (CPP).

Regulatory assistance project (RAP)   (Chapter 6) RAP is a global, nonprofit team of expert former regulators that focuses on the long-term economic and environmental sustainability of the power and natural gas sectors, providing technical and policy assistance to government officials on a broad range of energy and environmental issues. RAP has worked extensively in the United States since 1992, in China since 1999, and is expanding operations to Europe and India in 2011.

Remote terminal unit (RTU)   (Chapter 1) Microelectronic devices at the end of a communication system inside an electric utility, RTUs are connected to batteries and send and receive data and control signals as part of a SCADA system or DCS to enable energy control center operators to have visibility and control of multiple elements of the grid within a substation.

Renewable portfolio standard (RPS)   (Chapter 6) State legislative activity on smart grid is focused on renewable energy, specifically on setting RPSs. An RPS—RES at the federal level—is used at the state level to provide utilities with an achievable target in an achievable time frame, encouraging them to shift to a more sustainable energy portfolio over time. RPS is also used to encourage economic development and job growth based on clean energy. To date, 30 states and the District of Columbia have such standards. As a matter of policy, the American Wind Energy Association supports the adoption of a national RES, stressing the impact on jobs. RES and RPS have proven to be an important concept to drive the growth of the DG market, specifically rooftop solar PV, but many challenge the idea of a national RES, given the widely divergent environments and situations across the United States.

Request for information (RFI), request for quote (RFQ), request for proposal (RFP)   (Chapter 6) All three of these requests are formal procurement mechanisms for organizations. An RFI may be simply an attempt to identify qualified vendors and further information for a subsequent RFQ or RFP, or it may in fact become an actual purchasing mechanism by the buyer. An RFQ is stronger than an RFI, but less intentional than an RFP. An RFP is a serious commitment to purchase products and/or services, but if the buyer is dissatisfied with responses, it may still not result in a purchase. A standard process is for a short list of qualified vendors to be created from the most attractive respondents, after which due diligence and oral presentations may proceed before contract negotiations and the announcement of a winner.

Return on investment (ROI)   ROI is one of many terms used to describe the economic benefits that derive from a particular investment (others include IRR and ROE). The formula for ROI is (investment + gain)/investment, to produce a percentage number above 100% if the project is profitable. ROI is typically stated as a percentage, citing the original investment and the investment period, as in "the project provided an ROI of 150% on an original investment of $1 million over the three-year project life." ROI is a helpful tool in smart grid planning to assess the potential benefits of multiple projects and compare returns and relative risks.

RF mesh network   In a mesh network, client devices such as smart meters, access points, and other wireless devices communicate with each other, passing along a signal until it can be delivered to its destination, or in the other direction, to the internet at an internet gateway. Mesh radios talk to each other multiple times per second while they configure new data paths for optimal data transfer. Mesh network design is based on coverage area, topography, and anticipated data bandwidth requirements. RF mesh network nodes are complemented in a wide-area network by point-to-point radios and wired networks, because the technology requires a network design with sufficient numbers of "gateway nodes," which are used to provide backhaul—a direct connection to the internet, either through a wired connection—fiber, T-1, DSL, or cable—or a wireless point-to-point radio to an internet connection.

Roaming (Chapter 7) A general term in wireless telecommunications that refers to the extending of connectivity service in a network that is different than the network with which a station is registered. The primary example of "roaming" is for cell phones, when one takes one's phone to an area where the service provider does not have coverage (e.g., another country).

Second-generation smart grid (smart grid 2.0 or SG2)* (Chapter 1) A smart grid project that begins with a deliberate effort to craft a smart grid architecture framework and design, followed by an integrated IP network, which enables the seamless, subsequent addition of applications and end devices.

Service-level agreement (SLA) (Chapter 4) An SLA is a contractual mechanism that provides specific time or service levels and includes specific remedies when those standards are not met. The value of an outsourced service with an SLA compared to self-provisioning is that when difficulties are encountered, the outsourced agent must pay penalties, whereas an internal service has no such remedies.

Smart grid (Preface and Chapter 1) The integration of an electric grid, a communications network, software, and hardware to monitor, control, and manage the creation, distribution, storage, and consumption of energy. The smart grid of the future will reach every electric element. It will be self-healing, interactive, and distributed.

Smart grid 1.0 (SG1)* (Preface and Chapter 1) See *first-generation smart grid*.

Smart grid 2.0 (SG2)* (Preface and Chapter 1) See *second-generation smart grid*.

Smart grid 3.0 (SG3)* (Preface and Chapter 7) See *fast forward to smart grid 3.0*.

Smart grid architecture framework (SGAF).* (Chapter 7) An SGAF is a set of standards, best practices, rules, and methodologies to build a smart grid architecture—answering the how question. Smart grid architecture is comprised of infrastructure (networking, security, computers, and data storage systems), data, applications, and processes. The architecture design is the written plan and blueprint to direct activities, and the architecture framework (standards, best practices, rules, and methodologies) becomes the set of instructions on how to build the smart grid. The framework is the cookbook by which one builds a smart grid; the architecture itself is the artifact that describes the smart grid.

Smart Grid Consumer Coalition (SGCC) (Chapter 6) The SGCC is a nonprofit organization with an inclusive approach, gathering all stakeholders—consumer and environmental advocates, technology vendors, research scientists, and electric utilities—to listen, educate, and collaborate toward modernized electric systems. SGCC's focus on inclusive, consumer-driven change in the United States is a unique smart grid advocacy approach. SGCC urges industry leaders to pool resources to learn what consumers need and ways to communicate based on an understanding their concerns: listen via consumer research, educate via outreach and messaging tool kits, and collaborate via shared best practices.

Smart Grid Interoperability Panel (SGIP) (Chapter 2) NIST created SGIP in 2009 to engage smart grid stakeholders for technical assistance in assessing standards needs and developing the smart grid interoperability framework.

Smart grid optimization engine (SGOE)* (Chapter 7) The SGOE includes functionality similar to a traditional utility network-modeling tool but focuses on dynamic balancing of volt/VAR levels based on real-time data inputs from a multitude of devices. The SGOE manages the grid in a predictive manner, anticipating failures before they occur.

Solar Energy Industry Association (SEIA) (Chapter 6) SEIA represents the solar energy industry, which saw significant events develop in 2010, including a relentless march toward solar energy grid parity, as solar energy approaches become cost-competitive with fossil fuels, dropping to as low as $0.09–0.10 or less by 2020. Intermittency remains a challenge for solar PV and will become a greater challenge as PV market penetration increases.

Solar garden Smaller than a solar farm, solar gardens have gained popularity as "right-sized" energy systems suitable for cooperative sharing or for community energy purposes.

Solar photovoltaic (solar PV) (Chapter 2) See *photovoltaic*.

Standard energy efficiency rate (SEER) (Chapter 3) SEER is a way to rate the efficiency of air conditioners as defined by the Air Conditioning, Heating and Refrigeration Institute (ARI 210/240 is the SEER standard). A cooling unit's SEER rating equals cooling output in British thermal units during a typical cooling-season divided by the total electric energy input in watt-hours during the same period. A higher SEER rating means greater energy efficiency.

Supervisory control and data acquisition (SCADA) (Chapter 1) SCADA systems are used to provide remote access to conditions on a distributed infrastructure—for our purposes, SCADA systems provide utility control center operators visibility of grid conditions at the transmission and distribution levels. SCADA subsystems include (1) an HMI for operators to monitor and control processes; (2) a supervisory system (which, these days, means computers) to gather data on a process and send commands; (3) RTUs, which sit at the ends of the network and connect to sensors, to digitize sensor signals and send them to the supervisory system; (4) programmable logic controllers (plcs), more convenient and economical modern alternatives to RTUs; and (5) telecommunications infrastructure, wired or wireless, that connects the supervisory system to the field devices, either RTUs or PLCs. In electric utility grids, SCADA systems are most often seen in combination with EMSs, and in function, are similar to DCSs.

System integrator (SI) An SI brings together multiple pieces of a complex system such as a smart grid project, just as a general contractor takes responsibility to get a house built. The SI serves a vital role as a single point of contact responsible for the creation of the system.

T-1 A dedicated digital communication link provided by a telephone company that offers 1.5 megabits per second of bandwidth, commonly used for carrying traffic to and from private business networks and internet service providers.

Telecommunication Any transmission, emission, or reception of information of any nature by wire, radio, visual, or other electromagnetic systems.

Time-of-use (TOU) TOU rates use interval data from smart meters to bill customers as an alternative to flat rates. With TOU rates, the day is divided into periods—off-peak, shoulder, peak, shoulder is a fairly standard approach to TOU rate design, whose purpose is to create a significant difference between peak and off-peak rates to give customers a price incentive to consume more energy during off-peak periods and less during peak. A challenge of TOU ratemaking is to assign the appropriate delta between rate periods to drive the desired consumer behavior and ensure sufficient revenue for the utility.

Total cost of ownership (TCO) TCO is an approach to economic evaluation that includes a comprehensive assessment of system or project costs, for two or more projects with similar goals, so that an apples-to-apples comparison of costs may be made. Subsequent to a TCO analysis, benefits may be added to provide a business case comparison.

Transmission Control Protocol/Internet Protocol (TCP/IP)   TCP/IP is the set of standards for how computers, smart phones, and other digital devices communicate with each other over networks. Developed in the 1970s, TCP/IP allowed computers from different manufacturers to talk to each other in a common way for the first time, and TCP/IP became the foundation for communication on the Internet.

True sine wave (TSW)   In contrast to a modified sine wave, which has a squared off shape, a TSW closely resembles the natural curved sine wave found in physics and, more importantly, in generated electricity. While a modified sine wave is adequate for cheaper inverters and for basic uses, a TSW is preferred for grid-tied inverters used with solar PV systems and other DER devices, for a number of reasons. TSW inverters provide output voltage waveform with very low harmonic distortion and clean power like utility-supplied electricity and inductive loads (microwaves and motors) run faster, quieter, and cooler. TSW reduces audible and electrical noise in fans, fluorescent lights, audio amplifiers, TV, game consoles, faxes, and answering machines and prevents crashes in computers, weird printouts, and glitches and noise in monitors. Some appliances may not work with modified sine wave.

U.S. Green Building Council (USGBC)   (Chapter 6) The USGBC, a 501(c)(3) non-profit community of leaders, has a goal to make green buildings available to everyone within a generation. USGBC is perhaps best known for its signature LEED program (see LEED), which provides a roadmap and certification for buildings that are built according to "green" standards.

Utilities Telecommunications Council (UTC)   (Chapter 6) Having represented utility telecommunications issues and utilities since 1948, UTC has the perspective needed to contribute to the debate on a smart grid transition. From its Smart Grid Policy Summit in Washington, D.C., in April of 2010, to its online UTC Insights, the smart grid focus of UTC blends technology and policy expertise. In September 2010, UTC released a report sponsored by Verizon, titled "A Study of Utility Communications Needs: Key Factors That Impact Utility Communications Networks," which addressed a key challenge for utilities: deciding when it's appropriate to build and own a network and when it's better to subscribe to services delivered over a carrier network. The design of such hybrid networks will go a long way to the success of a smart grid project.

Validation, estimation, and editing (VEE)   A component of MDM, VEE is the process that makes raw data from smart meters suitable for digital billing and other utility operational purposes. The meter data must first be verified for accuracy (validated); then, when gaps are identified, estimates are created (estimated), and when faulty data are identified (edited).

Vehicle-to-grid (V2G), vehicle-to-home (V2H)   (Chapter 7) These two terms are used to describe the transfer of power from an EV to the grid or to the home, where the EV becomes an alternative power supply.

Virtual power plant (VPP)   (Chapter 7) VPP describes a demand-side alternative to accommodate growth in peak demand to the traditional supply-side alternative of adding a natural gas power plant, commonly referred to as a peaking unit or a peaker. In its most expansive definition, a VPP combines an array of rooftop

PV systems with localized ES and aggregated DR capacity (e.g., HEMS appliances equipped with direct load control—commonly, smart thermostats—or some combination). Such a system provides a utility the capacity needed to meet its peak needs without adding a power plant.

Virtual private network (VPN)   A VPN is virtual because it creates a private network by using a public network as the means of transporting the information. VPNs rely on encryption and other network security means to ensure that their private information is not intercepted while on the public network.

Voice over IP (VOIP)   VOIP is a relatively new type of telecommunication service that transmits digital signals using IP over the Internet rather than the PSTN—the traditional system of interconnected phone lines and switches.

Wide area network (WAN)   A WAN is a network that covers a broad area (i.e., any telecommunications network that links across metropolitan, regional, national, or international boundaries) using either privately owned or leased telecommunication lines. Business and government entities utilize WANs to relay data among employees, clients, buyers, and suppliers from various geographical locations. In essence, this mode of telecommunication allows a business to effectively carry out its daily function regardless of location. The Internet can be considered a WAN as well and is used by businesses, governments, organizations, and individuals for almost any purpose imaginable.

Wi-Fi   Wi-Fi is a brand name created by the Wi-Fi Alliance and attached to a set of global technology standards for wireless data networking equipment. Wi-Fi® is a registered trademark of the Wi-Fi Alliance, a nonprofit international association formed in 1999 to certify interoperability of wireless LAN products based on IEEE 802.11 specifications. Headquartered in Austin, Texas, the Wi-Fi Alliance is comprised of member companies from around the world, and has certified thousands of products since certification began in March of 2000. When the IEEE standards committee agreed to a set of standards, they allowed manufacturers to produce equipment that was interoperable. Wi-Fi has been widely applied in enterprises to expand wired LANs and enhanced to create RF mesh networks for citywide broadband. Wi-Fi LANs have a typical range of 100–500 feet. 802.11b, the original Wi-Fi standard, operates at up to 11 Mbps in the 2.4-GHz band. 802.11g operates at up to 54 Mbps in the same 2.4-GHz band as 802.11b and is backward-compatible with 802.11b access points. 802.11a operates at up to 54 Mbps in the 5-GHz band and is not backward-compatible with 802.11b. 802.11n is the most recent standard to further increase the throughput of wireless networks, providing greater than 100 Mbps usable throughput over an 802.11 communications channel.

WiMAX   (Chapter 1). WiMAX is a trademark of the WiMAX Forum. Also known as IEEE 802.16, WiMAX stands for Worldwide Interoperability for Microwave Access, which is a point-to-multi-point, non-line-of-sight (NLOS) wireless broadband access technology. WiMAX can transfer data at rates around 70 Mbps with a range of 10–30 miles. WiMAX base stations can cover an entire metropolitan area, making that area into a WMAN and allowing true wireless mobility within the coverage area. The WiMAX standard relies mainly on spectrum in the 2–11 GHz range. The WiMAX specification was created to improve upon many of the limitations of the Wi-Fi standard by providing increased bandwidth and providing stronger encryption.

Wireless   A broad term used to describe any telecommunications wherein the signal moves through the air, not through a wire. The breadth of the term can cause confusion: Originally, "wireless" was a synonym for radio; in more recent years, "wireless phones" referred specifically to mostly voice communication over cellular networks; and in the last few years, "wireless data" has come to refer also to newer data technologies such as Wi-Fi, WiMAX, and LTE.

Wireless local area network (WLAN)   A term used to describe a small wireless network, synonymous with hot spots and access points.

Wireless metropolitan area network (WMAN)   A large network that covers a broad metropolitan area such as an entire major city. Historically operated by major wireless carriers, WMANs provide voice and relatively low-bandwidth data service over a broad area and require a dense network of communications towers, as well as spectrum licenses from the government. WANs can cost hundreds of millions of dollars to establish and maintain.

ZigBee   (Chapter 1) ZigBee is an industry specification for a communication protocol using small, low-power digital radios for short-range communication, based on the IEE 802.15.4-2003 standard. In the smart grid industry, ZigBee has been widely adopted by smart meter manufacturers to communicate meter data to IHDs.

About the Authors

Andres Carvallo is the CEO of CMG, a strategy consulting and advisory company enabling smarter cities, utilities, enterprises, vendors, and start-ups.

Andres, who is globally recognized by the IEEE as one of the early developers of the smart grid concept and technology, defined the term smart grid on March 5, 2004. Andres, as its CIO, championed Austin Energy's industry leading smart grid program design and implementation from 2003 to 2010, and he architected the Pecan Street Project from 2008 to 2010. Furthermore, Andres co-led the investments in many green generation facilities, telecom infrastructure and corporate systems, while leading a well-known and successful technology and process transformation company-wide. For his successes, Andres has received 34 industry awards since 2005 and is a popular speaker and guest lecturer.

Andres has over 28 years of experience in the energy, telecommunications, computer, and software industries. Since 1992, Andres has held P&L responsibilities and senior executive titles while being responsible for the strategy, development, and commercialization of over 40 products at start-ups like Proximetry, Grid Net, HillCast, AGEA (agentgo), iMark.com, and Tycho Networks and leading global companies like Philips Electronics, Digital Equipment, and Borland. Furthermore, Andres started his career as a Windows product manager at Microsoft.

Andres received a B.S. in mechanical engineering degree from the University of Kansas with a concentration in robotics and control systems. Andres has also postgraduate certificates in business management from Stanford University, in quality management from the Wharton School of the University of Pennsylvania, and in power utility management from the University of Idaho.

John Cooper is a creative thinker, author, researcher, change agent, and project manager, active in the energy, telecommunications, IT services, and government research industries since the mid 1980s. With extensive experience in business development and consulting positions at innovative electric and telecommunications companies, John has been responsible for leading innovation projects in all aspects of the emerging smart grid, ranging from utility IP networks, wireless advanced metering infrastructure, DG, DR, EE, utility-scale energy storage, VPPs, EV charging infrastructure, smart grid roadmapping, and business transformation. John currently works with Siemens' new Business Transformation Services division, helping utilities plan and execute a managed transition into a sustainable energy business based on new utility business and industry models. Prior, John served as smart grid subject matter expert in projects in New Brunswick, Canada, and Ankara,

Turkey. John was previously with NextWatt Solutions (commercial building DER solutions), UtiliPoint (smart grid industry analyst), and Grid Net (vice president for utility solutions). John provided consulting services as the president of his consulting firms Ecomergence and MetroNetIQ, where notable clients included Sharp Labs of America (research on VPP business models), the Pecan Street Project (smart grid project manager and author of the technical report), Xtreme Power (utility-scale ES company), GRIDbot (EV charging start-up), various clients regarding multipurpose broadband networks for AMI and other applications, and Austin Energy (pioneer GENie Project in 2004). From 2000 to 2002, John helped to launch Dell's entry into managed services while at Dutch global service provider Getronics. In the mid 1990s, John worked with Central and South West, an IOU now merged with AEP, to plan one of the first wireless AMR projects in Tulsa, Oklahoma. Early in his career, John helped launch and then served as the director of the Texas Senate Research Center. John is also the author of The ABCs of Community Broadband, a smart cities guide for community leaders, as well as numerous white papers and magazine articles on smart grid. John received a B.A. in government from the University of Texas at Austin and an M.B.A. with honors from the McCombs School of Business at the University of Texas at Austin.

Index

CPSIA information can be obtained
at www.ICGtesting.com
Printed in the USA
JSHW060823250722
28253JS00002B/125